# Nuclear Engineering Division



Presentations at the 2009 AIChE Annual Meeting

## Nashville, Tennessee, USA
## 8-13 November 2009

ISBN: 978-1-61567-936-2

**Printed from e-media with permission by:**

Curran Associates, Inc.
57 Morehouse Lane
Red Hook, NY 12571
www.proceedings.com

**Some format issues inherent in the e-media version may also appear in this print version.**



Printed by Curran Associates, Inc. (2010)



AIChE
3 Park Avenue
New York, NY 10016-5991

Phone: (203) 702-7660
Fax: (203) 775-5177

www.aiche.org

# TABLE OF CONTENTS

# Shaping Our Energy Future: Advanced Research at the U.S. Department of Energy

**Jacques Beaudry-Losique**, Department of Energy, Washington, DC

The American Recovery and Reinvestment Act of 2009 provided more than $60 billion in clean energy investments. Building upon its record of successful research and development, the Department of Energy is focusing its efforts on meeting the energy objectives of the Administration, including new job creation, breaking our dependence on oil, increasing domestic energy production, promoting energy efficiency, closing the carbon loop-hole, and ensuring U.S. competiveness. This presentation will discuss research and development as well as partnership activities the DOE Office of Energy Efficiency and Renewable Energy is performing to deliver diverse, clean, reliable and affordable energy technologies to the marketplace.

See more of this Session: 2009 Annual Meeting Plenary Session - Looking Forward - Energy Policy & Technology
See more of this Group/Topical: Liaison Functions

# The Challenges of Scale and Sustainability in U.S. Energy Policy

**Marilyn Brown**, Public Policy, Georgia Institute of Technology, Atlanta, GA

The pace of technology improvement and deployment could significantly reduce the cost of stabilizing greenhouse gas concentrations and improve the nation's energy security. Given the need for a large-scale energy transformation, the challenge is to move toward actions that go beyond technology R&D to strategies that target the rapid and large-scale absorption of clean energy technologies into the economy. However, there are numerous barriers to the commercialization and deployment of technology innovations; those that are imposed by administrators, legislatures, and regulators are particularly of interest as they operate at cross-purposes with the U.S. government's GHG reduction goals. This talk focuses on these public policy issues, particularly the challenges of scale and sustainability.

See more of this Session: 2009 Annual Meeting Plenary Session - Looking Forward - Energy Policy & Technology
See more of this Group/Topical: Liaison Functions

# The Nation's Renaissance in Energy Technology

**Dana C. Christensen**, Energy & Engineering Sciences Directorate, Oakridge National Laboratory, Oak Ridge, TN

The nation is facing a renaissance in energy technology to include dramatic changes in transportation, electricity production, distribution and use, and energy efficiency. The nation is also rapidly expanding its scientific capabilities and is currently leading the world in areas such as material science, characterization capabilities, and computational science. Today's industrial challenges are increasingly complex and require the integrated use of these tools in order to solve the critical problems in an economical fashion. In fact, it is becoming increasingly clear that science and technology today requires an integrated approach that includes theory, experimentation and computational science. This new three legged stool for science will be discussed and a number of examples involving the changes in energy technology will be presented.

See more of this Session: 2009 Annual Meeting Plenary Session - Looking Forward - Energy Policy & Technology
See more of this Group/Topical: Liaison Functions

# Clean Coal Technology and CO2 Capture: A Perspective

**L.-S. Fan**, Chemical and Biomolecular Engineering, The Ohio State University, Columbus, OH

Absolute and per-capita energy consumption is bound to increase globally, leading to a projected increase in energy requirements of 50% by 2020. The primary source for providing a majority of the energy will continue to be fossil fuels. However, an array of enabling technologies needs to be proven for the realization of a zero emission power, fuel or chemical plants in the near future. Opportunities to develop new processes, driven by the regulatory requirements for the reduction or elimination of gaseous and particulate pollutant abound.

This presentation will describe the chemistry, reaction mechanisms, particle technology, system engineering, process economics, and regulations that surround the modern utilization of fossil energy, particularly, for the clean coal technology. The presentation will illustrate the salient features pertaining to the fundamental and applied characteristics of the state-of-the-art technologies in practice as well as emerging technologies in development. Emerging technologies evolve largely from the urge for CO2 emission control in fossil energy conversion systems. Further, novel gasification systems based on the chemical looping concepts as well as the calcium looping process for CO2 separation from the combustion flue gas stream will be elucidated in the context of the looping particle design, process heat integration, energy conversion efficiency and economics.

See more of this Session: 2009 Annual Meeting Plenary Session - Looking Forward - Energy Policy & Technology
See more of this Group/Topical: Liaison Functions

# Advanced Biofuels - The Path Forward

**W. Densmore Hunter**, Catchlight Energy, LLC, Federal Way, WA

Oil prices go up and they come down and our ability to forecast them with any real confidence has not been particularly good. A year ago when oil prices were at record highs and global economies were booming, the case for biofuels seemed clear. The path forward is cloudier today. Are we facing a repeat of the boom and bust cycle that decimated alternative energy development in the 1980s? What are the important differences to prevent another collapse? Four factors - climate change, rising global demand for a constrained supply, the end of easy oil, and the political and economic imperatives to reduce oil imports - are expected to sustain biofuels development, despite the hurdles brought about by the global economic crisis. It's a very different situation from the early 1980s. The push for sustainable energy is real and it's here to stay.

How will advanced (lignocellulosic) biofuels develop? It is a much more technically demanding challenge and it will take longer to unfold than grain based ethanol. While chemically similar as polymers of simple sugars, there are important differences between starch and cellulose. Those differences manifest in the way we have traditionally used them. We eat starch. We build houses out of cellulose. Processes involving hydrolysis and fermentation face far greater hurdles with lignocellulosic feedstocks than with starch. In fact it's quite likely that no single technology will dominate the 2nd generation biofuels industry. There are many potential pathways involving biochemical, thermochemical and chemical processes. It's too early to pick the winners. Different situations will favor different solutions. The speaker will explore promising developments that are likely to shape the emerging biofuels industry together with some of the challenges and uncertainties. One conclusion that stands out clearly among all the uncertainty is that the chemical engineering profession will play a dominant role in making sustainable biofuels a reality and thereby contributing in a significant way to the resolution of critical global issues.

See more of this Session: 2009 Annual Meeting Plenary Session - Looking Forward - Energy Policy & Technology
See more of this Group/Topical: Liaison Functions

# Commercial Scale Tritium Recovery From Exit Signs and Other Illumination Devices at LLNL

Diane Spencer, Weapons and Complex Integration Directory, Lawrence Livermore National Laboratory, Livermore, CA

J. Mark Mintz, Weapons and Complex Integration Directorate, Lawrence Livermore National Laboratory, Livermore, CA

The disposal of exit signs and similar commercially available tritium-powered illumination devices (TPIDs) in landfills can pose potential public health and environment risks. Studies of active landfills in various states and countries show leachate with tritium considerably above background, with the tritium concentration in many above the Environmental Protection Agency's drinking water MCL of 20,000 pCi/L and between 100,000 pCi/L and 200,000 pCi/L in others. The environmental contamination risk posed by improper disposal of obsolete TPIDs can be greatly reduced by collecting the gaseous tritium remaining in TIPDs. Beginning 2001, LLNL and the U.S. Army have been engaged in a successful tritium recovery and recycle program for a limited number of their specialized devices. This laboratory-scale work has been accomplished by hand with reasonable efficiency. Recent requests for assistance with similar work, however, include unit volumes in the millions and require much larger capabilities. As a result, the LLNL Tritium Facility has installed and will soon begin operating a commercial-scale tritium recovery system-the Tritium Grinder System (TGS)-to complement its laboratory-scale tritium recovery capabilities. The TGS can process up to 72 standard exit signs in a batch, with significantly more for smaller illumination devices (e.g., 2250 rotary phone dials per batch). The "grinder" portion of the TGS is a commercially available shredder unit from SSI Shredding Systems, Inc., modified to operate under vacuum. A TIPD batch is delivered to the grinder by conveyor belt and is mechanically shredded. The released tritium, along with decay product 3He and Ar added as a sweep gas, is then vacuum-pumped to a collection tank. The collected gas is processed through a regenerable hydrogen getter to remove the tritium product. Solid debris (e.g., exit sign housing residuals) generated through the recovery process is collected in a lined 55-gallon drum using a "bagout" process engineered to minimize outgassing and worker exposure. Upon completion of the bagout process, the content of the drum is ready for final disposal as low-level radioactive waste.

See more of this Session: Chemical Engineering Advances in the Processing of Radioactive Waste
See more of this Group/Topical: Nuclear Engineering Division

# CHARACTERIZATION OF INDIVIDUAL CHEMICAL REACTIONS CONSUMING ACID DURING NUCLEAR WASTE PROCESSING AT THE SAVANNAH RIVER SITE – 136b

*David C. Koopman, Savannah River National Laboratory, Aiken, SC*
*Bradley R. Pickenheim, Savannah River National Laboratory, Aiken, SC*
*Daniel P. Lambert, Savannah River National Laboratory, Aiken, SC*
*J. David Newell, Savannah River National Laboratory, Aiken, SC*
*Michael E. Stone, Savannah River National Laboratory, Aiken, SC*

## ABSTRACT

Conversion of legacy radioactive high-level waste at the Savannah River Site into a stable glass waste form involves a chemical pretreatment process to prepare the waste for vitrification. Waste slurry is treated with nitric and formic acids to achieve certain goals. The total quantity of acid added to a batch of waste slurry is constrained by the catalytic activity of trace noble metal fission products in the waste that can convert formic acid into hydrogen gas at many hundreds of times the radiolytic hydrogen generation rate. A large block of experimental process simulations were performed to characterize the chemical reactions that consume acid prior to hydrogen generation. The analysis led to a new equation for predicting the quantity of acid required to process a given volume of waste slurry.

## INTRODUCTION

Legacy radioactive high-level waste (HLW) generated at the U. S. Department of Energy - Savannah River Site (SRS) during production of enriched uranium and plutonium during the Cold War is currently being processed in the Defense Waste Processing Facility (DWPF) into a stable borosilicate glass waste form for long term storage. The majority of the legacy HLW is stored either as mixtures of hydroxides and hydrous oxide insoluble solids or as precipitated salt cakes in large cylindrical storage tanks at SRS. These 3.5-4.9 thousand cubic meter (900,000-1,300,000 gallon) carbon steel storage tanks also contain 5-7M sodium solutions rich in hydroxide, nitrate, and nitrite anions. Batch feed preparation for processing in the DWPF involves washing the aqueous phase of slurries obtained from one or two individual waste tanks to about a 1M sodium concentration.

The DWPF brings approximately 25 cubic meters of fresh washed sludge slurry into the Sludge Receipt and Adjustment Tank (SRAT) at a time. The SRAT is the primary vessel for chemical pretreatment of the HLW prior to vitrification. The DWPF primarily employs batch processing, although the waste glass melter is fed semi-continuously. Nitric and formic acids are added during the SRAT cycle to adjust the rheology, dissolve some of the alkaline earth and transition metal compounds, convert HgO to elemental mercury for steam stripping, destroy some of the existing anions such as carbonate and nitrite, and chemically reduce a portion of the $MnO_2$ to $Mn^{2+}$. (Thermal reduction of Mn in the DWPF melter leads to reduced melter throughput and can also lead to foam formation.) Optimal performance of the DWPF waste glass melter requires a balance of chemical oxidizers and reductants in the feed in order to avoid foam formation and metal precipitation. The balance is achieved by dividing the SRAT acid demand between nitric acid (oxidizer) and formic acid (reductant).

The original concept for the DWPF involved a dual feed capability that would process insoluble radioactive waste solids in parallel with soluble radioactive salt waste cake material. The DWPF flowsheet was subsequently modified to substitute 90 wt% formic acid for the dilute formic acid stream coming from salt waste processing. A stoichiometric acid requirement equation was derived based on what was known at the time about the process chemistry, equation [1]:

$$\frac{\text{moles acid}}{\text{L slurry}} = \text{base equivalents} + 2 * \text{TIC} + 0.75 * \text{nitrite} + 1.2 * \text{Mn} + \text{Hg}$$

Equation 1, current estimating equation for the stoichiometric acid requirement.

All right-hand-side quantities in eqn. [1] are measured for the slurry (aqueous phase plus insoluble solids) and converted to moles per liter slurry units. Each term is measured independently (five different analytical instruments). The DWPF uses averages over four independent samples for each analysis. The base equivalents term represents the equivalent hydroxide molarity of the slurry determined by a strong acid titration to pH 7. The total inorganic carbon (TIC) term represents the total carbonate content of the slurry.

This approach to predicting total acid requirements was initially seen as an interim measure that would not be needed once "short-term" issues with salt waste processing were resolved. The DWPF, however, continues to operate in the interim mode fourteen years later. Supporting experimental work done during that time has shown that the above equation predicts the acid demand 20-30% lower than the observed demand. An empirical multiplier has been used as a correction factor for each new large batch of feed. Each large sludge batch typically is used to provide feed for 70-100 SRAT cycles during which the multiplier is typically held constant.

Changes in the preparation strategy of the legacy waste sludges over the past decade have been made to reduce the volume of decanted wash water, since tank farm storage space and evaporator capacity are limited. The resulting DWPF feed sludge compositions require proportionally more acid to process (increased hydroxide, carbonate, and nitrite ion concentrations in the DWPF feed). This doctrinal evolution has constrained the quantity of acid that can be added relative to the minimum required to accomplish various processing goals that constituted the original design concept.

Tightened constraints have renewed interest in understanding the nature and timing of the chemical reactions that consume acid. Predicting and controlling the quantity of acid added in the SRAT is an essential part of the DWPF operational strategy for minimizing noble metal catalyzed hydrogen generation from excess formic acid. Alternative salt processing has started at SRS. Streams concentrated in radionuclides from salt separations are now being incorporated into the DWPF flowsheet. These streams are typically either acidic or basic, and consequently they impact the actual acid requirement. The quantity of salt processing streams in a given SRAT cycle varies considerably. This causes increased variability in the acid requirement. The variability in actual acid demand raises the issue of whether or not a single multiplicative correction factor is adequate to handle a wide range of feeds.

Another issue with the acid equation (despite the fact the equation predicted low) was the potential for double counting certain species. For example, since measurement of hydroxide ion concentration was problematic in the system, a slurry titration to pH 7 was substituted. A total inorganic carbon measurement was used to get the carbonate concentration of the slurry. Soluble carbonate, however, was partially titrated to bicarbonate during the titration measurement and was being counted twice for acid demand. Experiments showed that it was only soluble carbonate, and not insoluble carbonate, that was being titrated and counted twice in the current equation. There was also a concern that HgO might be dissolved during titration. Experiments indicate that mercury was not appreciably dissolved during pH 7 titrations, although titrations to pH 5.5 were being impacted.

Two other findings emerged from preliminary experiments. Different titration techniques produced different equivalent base molarities for the slurry. The DWPF used 20:1 to 25:1 diluted slurry in their automatic titrator, while some work within the Savannah River National Laboratory used direct titrations. Results by the two methods were typically within 5-15%. This has been deemed tolerable, although this term is often the largest contributor to the total in eqn. [1]. Obtaining a reliable measurement for slurry total inorganic carbon, however, has been a bigger issue. Uncertainties of reported quantities have been high. Therefore, finding methods to predict the stoichiometric acid requirement that did not require the measurement of slurry TIC or the titration to pH 7 were promoted over a strategy that only sought to refine some of the coefficients in the existing acid equation.

## TEST PROGRAM

A matrix of large lab-scale tests were completed in duplicate with an integrated sampling plan that permitted the chemistry of the SRAT during acid addition to be elucidated and better quantified. Batches of simulated waste were processed 15 $dm^3$ (15-L) at a time. Data from these tests filled gaps in the historical database. Reactions were identified that were actively consuming acid. The dissolutions of $Mg(OH)_2$ and $CaCO_3$ were successfully tracked as functions of pH during processing. These reactions were found to occur below pH 7 and at rates comparable to half of the acid addition rate (not unexpected, since $Mg^{2+}$ and $Ca^{2+}$ need two $H^+$ to neutralize their anions). They also were contributing significantly to the total demand for acid.

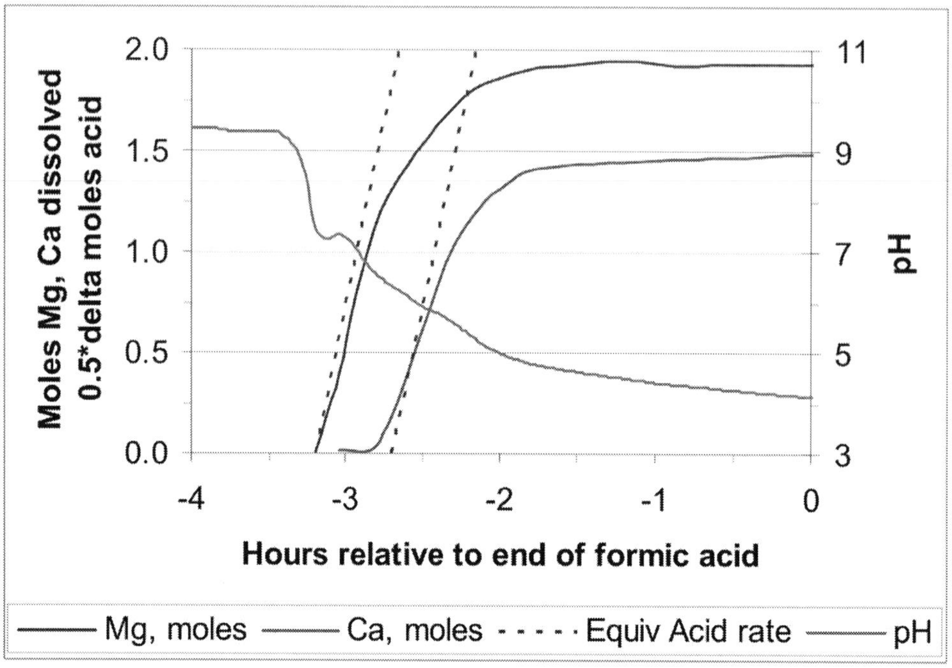


Figure 1. Dissolution rates of insoluble Mg and Ca during acid addition

The dissolution rate of Mg was initially nearly identical to the equivalent addition rate of acid protons. Mg is suspected to be present as $Mg(OH)_2$ with perhaps a small fraction of $MgCO_3$. The onset of $CaCO_3$ dissolution about two-thirds of the way through Mg dissolution occurred with a reduction in the rate of Mg dissolution. The combined Ca plus Mg dissolution rate was also essentially equal to the equivalent addition rate of acid protons. It is believed that some of the acid protons are reacting with soluble bicarbonate from the $Na_2CO_3$ in the starting aqueous phase during this same period. Total slurry TIC versus time data indicate that all carbonate was destroyed by the end of Ca dissolution. Integration of the carbon dioxide production rate through the end of Ca dissolution gave a value essentially identical to the moles of starting carbonate in the slurry feed.

The period following Ca and Mg dissolution was dominated by two main reactions. These were the reduction of Mn(IV) as insoluble $MnO_2$ to soluble Mn(II) and the destruction of nitrite ions, $NO_2^-$. Nitrite ions are destroyed by several parallel chemical pathways. Rh and Pd, present at 10-100 ppm in the waste from thermal neutron fission of $^{235}U$, catalyze two of these pathways. The main by-products are nitrate ion, NO, and $N_2O$. Since the SRAT is air-purged, most of the NO is converted to $NO_2$ (along with $N_2O_4$). Brown colored gas formation is observed during nitrite destruction. Gas chromatograph data indicate formation of NO and $N_2O$, and depletion of $O_2$. Ion chromatography tracks formation of nitrate ion during this period. The sequence of these reactions relative to Ca dissolution is shown in Figure 2.

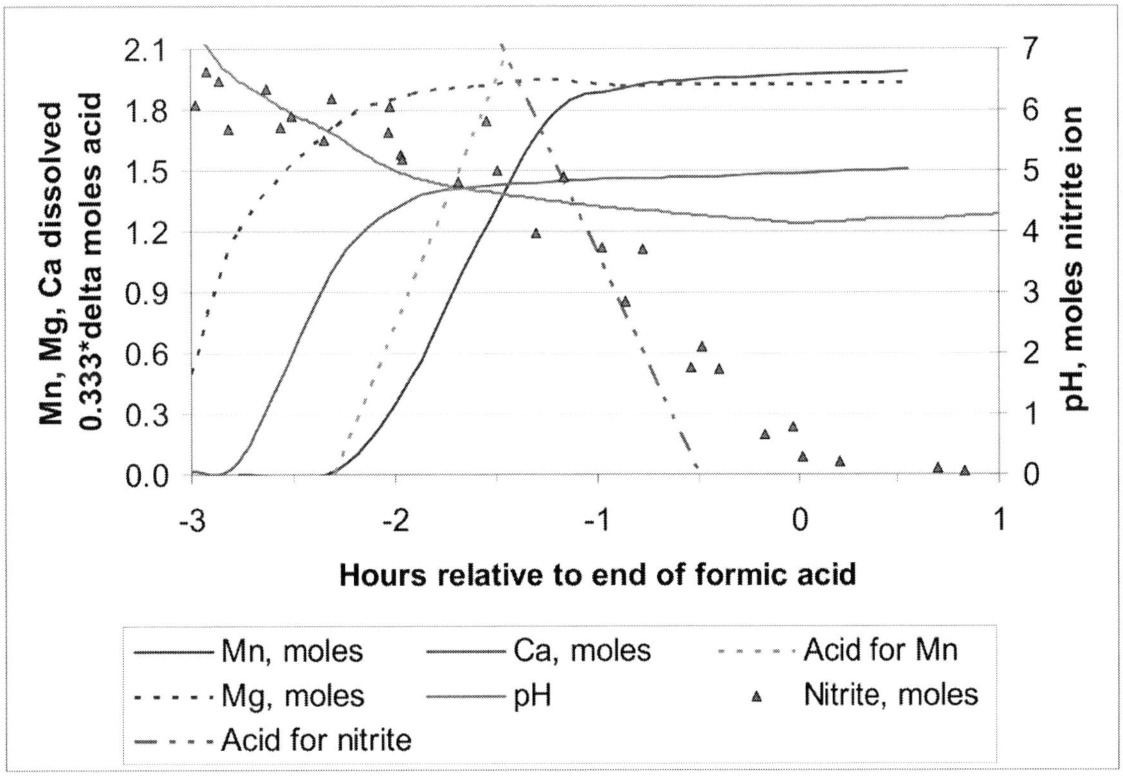


Figure 2. Mn reduction-dissolution and nitrite destruction after Ca dissolution.

$MnO_2$ reduction by formic acid requires consumption of three moles of acid per mole of Mn reduced.

$$MnO_2 + 3HCO_2H \rightarrow Mn(HCO_2)_2 + 2H_2O + CO_2$$

Figure 2 indicates that the rate of Mn dissolution is slower than one-third the rate of acid addition. This is partly due to the end of Ca dissolution and partly due to the onset of nitrite destruction. About half of the nitrite was destroyed during Mn reduction and the other half after Mn reduction. The reductive destruction of nitrite by formic acid also produces carbon dioxide. Consequently $CO_2$ formation is not directly useful for discriminating between the timing of Mn reduction and nitrite destruction. The mean consumption of acid by nitrite ion destruction is in the range of 0.9-1.5 moles acid per mole nitrite depending on catalytic activity. The amount of acid consumed increases with increasing noble metal concentrations. Typical HLW noble metal concentrations, however, generally result in about 1.0-1.1 mole acid per mole of nitrite destroyed.

Other reactions were found to consume acid during acid addition, but the initial concentrations and/or extent of reaction achieved were too small to have a significant impact on the calculated acid consumption. Some of these reactions included the partial dissolution of $Ni(OH)_2$, ZnO, and CuO. These reactions started once the pH was about five.

The following second generation stoichiometric acid equation was proposed to explain the data:

$$\frac{\text{moles acid}}{\text{L slurry}} = \text{base equivalents} + \text{soluble TIC} + \text{nitrite} + 1.5 * \text{Mn} + 1.5 * (\text{Ca} + \text{Mg}) + \text{Hg}$$

Equation 2, second generation equation for predicting the stoichiometric acid requirement in the DWPF SRAT cycle.

Soluble carbonate was given one mole acid per mole, since titration accounts for the second mole needed to convert carbonate to carbon dioxide. Calcium and magnesium were brought in at 75% of the stoichiometric requirement for $Mg(OH)_2$ and $CaCO_3$, since there exists a certain fraction of non-acid reacting compounds, such as oxalate, phosphate, sulfate, etc. associated especially with calcium and perhaps magnesium. These two terms are appear to account for the insoluble carbonate contribution so no measurement of slurry TIC is required. Elimination of the problematic analysis is considered a positive feature of the proposed equation.

Although the case shown in Figure 2 has Mn reduction essentially 100% complete prior to nitrite destruction, other simulations have produced contradictory results. The data in Figure 2 are from tests with low catalytic activity. Nitrite is often sufficiently destroyed prior to complete Mn reduction under other test compositions. The second generation stoichiometric acid equation has treated Mn as 50% reduced by complete nitrite ion destruction as a temporary compromise pending a more detailed evaluation of the available kinetic data.

Preliminary validation experiments show that the second generation equation fits the historical database much better than the original equation, while still retaining the first-principles premise that the chemistry can be understood in detail as a sum of individual reactions. Experimentally determined minimum acid requirements for nitrite destruction have been collected over the years and put into a statistical database for analysis using JMP statistical software. A simple comparison of the second generation equation to the minimum acid data is given in Figure 3.

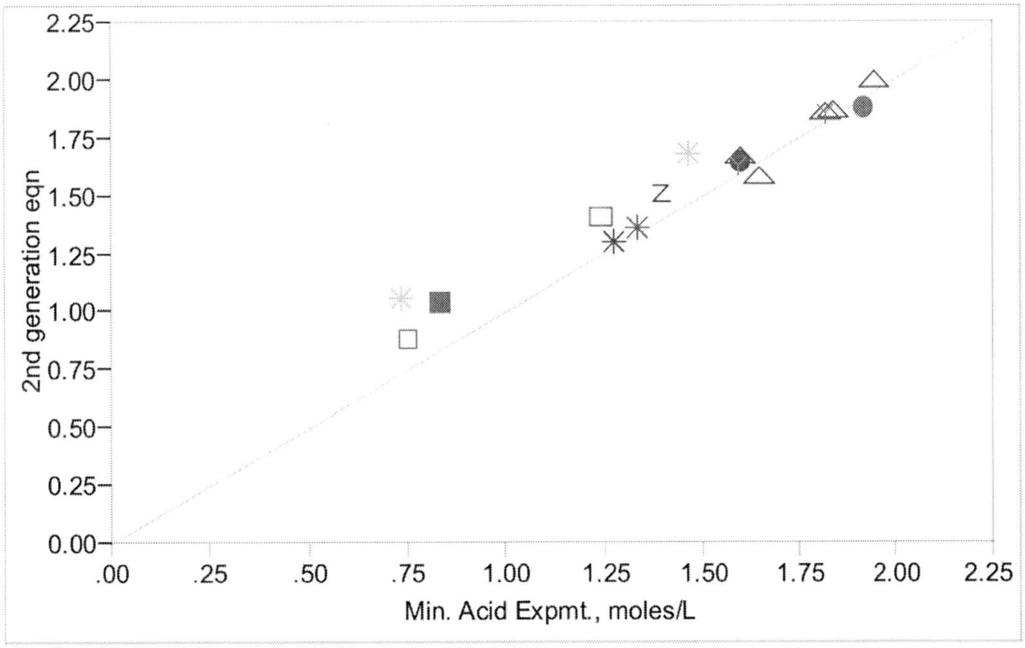


Figure 3. Validation of second generation stoichiometric acid equation predictions against experimentally determined minimum acid requirements.

The second generation acid equation shows a slight bias toward higher than measured values for acid demand but it generally matches the observed trend. Comparable agreement with successful plant operational data was obtained. Further studies are planned to investigate certain individual coefficients in the second generation equation and their impact on the fit of the equation to the existing data.

A second approach was taken toward evaluating the stoichiometric acid demand during SRAT processing. This approach led to an equation that contained neither titration nor TIC measurements. Consequently, it is not simply another evolution of the original equation (which is primarily driven by slurry anions). The premise was that a sum could be made over the soluble cations present after acid addition/consumption that would represent the cations that had had a basic anion neutralized by $H^+$. It was recognized immediately that this sum would overstate the acid demand, since not all of the associated anions were bases. Therefore, a credit for non-acid reacting anions was taken to reduce the result obtained by summing the soluble cations. The resulting equation is given as eqn. [3].

$$\frac{moles\ acid}{L\ slurry} = Na + K + Hg + 2*(Ca + Mg) + 1.5*Mn - NO_3^- - 2*SO_4^{2-} - HCO_2^- -$$
$$2*C_2O_4^{2-} - 3*PO_4^{3-} - Cl^- - F^- - NO_2^-*(fraction\ NO_2^-\ converted\ to\ NO_3^-)$$

Equation 3, cation-based prediction of the stoichiometric acid requirement.

The cation-based equation, like the two equations preceding it, has all terms expressed in terms of moles species per liter ($dm^3$) of waste slurry charged to the SRAT. The cation equation and the second generation equation track each other more closely than the original equation [1]. Both new equations seem to respond more appropriately to compositional changes than the original stoichiometric acid equation.

The cation-based stoichiometric acid equation appears to be conceptually sound. It would also eliminate two of the more difficult measurements in the original equation: base equivalents of the slurry and total slurry inorganic carbon. The main issues with the cation equation appear to be the large number of terms (14), the accumulation of measurement uncertainty (random error propagation), and the presence of both positive and negative terms. For example, the sodium term can be comparable to the entire stoichiometric acid requirement but is only accurate to about +/-10% when measured by inductively coupled plasma-atomic emission spectroscopy.

The new data on the chemistry of the major species present during SRAT processing have led to a greatly improved understanding of the sequence of reactions occurring and their impact on acid consumption. This has led to the development of two new equations for predicting the stoichiometric acid requirement for SRAT processing. The two new equations have already been implemented for lab-scale testing in parallel with the original equation. The plan is to attempt to implement a change in DWPF at the time of the next major sludge batch (within six months). At present, it will be the second generation stoichiometric acid equation that is implemented, since it has shown less random variability during validation tests than the cation equation.

# MERCURY-NITRITE-RHODIUM-RUTHENIUM INTERACTIONS IN NOBLE METAL CATALYZED HYDROGEN GENERATION FROM FORMIC ACID DURING NUCLEAR WASTE PROCESSING AT THE SAVANNAH RIVER SITE – 136c

*David C. Koopman, Savannah River National Laboratory, Aiken, SC*
*Bradley R. Pickenheim, Savannah River National Laboratory, Aiken, SC*
*Daniel P. Lambert, Savannah River National Laboratory, Aiken, SC*
*J. David Newell, Savannah River National Laboratory, Aiken, SC*
*Michael E. Stone, Savannah River National Laboratory, Aiken, SC*

## ABSTRACT

Chemical pre-treatment of radioactive waste at the Savannah River Site is performed to prepare the waste for vitrification into a stable waste glass form. During pre-treatment, compounds in the waste become catalytically active. Mercury, rhodium, and palladium become active for nitrite destruction by formic acid, while rhodium and ruthenium become active for catalytic conversion of formic acid into hydrogen and carbon dioxide. Nitrite ion is present during the maximum activity of rhodium, but is consumed prior to the activation of ruthenium. Catalytic hydrogen generation during pre-treatment can exceed radiolytic hydrogen generation by several orders of magnitude. Palladium and mercury impact the maximum catalytic hydrogen generation rates of rhodium and ruthenium by altering the kinetics of nitrite ion decomposition. New data are presented that illustrate the interactions of these various species.

## INTRODUCTION

Legacy radioactive high-level waste (HLW) generated at the Savannah River Site (SRS) during production of enriched uranium and plutonium during the Cold War is currently being processed in the Defense Waste Processing Facility (DWPF) into a stable borosilicate glass waste form for long term storage. Generally radionuclides account for less than 10% of the solid waste mass. Hydrous oxides of iron, aluminum, manganese, nickel and calcium are some of the main components. The majority of the legacy waste is stored in large cylindrical storage tanks at SRS. These 3.5-4.9 thousand cubic meter (900,000-1,300,000 gallon) carbon steel storage tanks also contain 5-7M sodium solutions rich in hydroxide, nitrate, and nitrite anions. Most SRS tank farm processes involve the transport of two phase slurries between tanks and the decanting of excess aqueous phases for processing in the tank farm evaporators. Batch feed preparation for DWPF involves washing the aqueous phase of slurries obtained from one or two waste tanks to about a 1M sodium concentration. A DWPF feed batch is typically about 1-2 thousand cubic meters of washed slurry.

The DWPF brings the washed sludge slurry into the Sludge Receipt and Adjustment Tank (SRAT) to begin batch processing. About 25 $dm^3$ are processed at a time. The SRAT is the primary chemical pretreatment vessel in the DWPF. The SRAT is operated like a batch reactor. Acids are added to adjust the rheology, dissolve some of the alkaline earth and transition metal compounds, convert HgO to elemental mercury for steam stripping, destroy some of the existing anions such as carbonate and nitrite, and chemically reduce a portion of the $MnO_2$ to $Mn^{2+}$. (Chemical reduction is desirable since thermal reduction of Hg and Mn in the DWPF melter can lead to reduced melter throughput foam formation.) Optimal performance of the DWPF waste glass melter requires a balance of chemical oxidizers and reductants in the feed in order to avoid foaming and prevent precipitation of metals. The balance is achieved by dividing the SRAT acid demand between nitric acid (oxidizer) and formic acid (reductant).

The legacy HLW contains small concentrations (1-200 ppm) of noble metals formed from $^{235}U$ thermal neutron fission product decay chains. In particular, the HLW contains Ru, Rh, and Pd in the approximate relative proportions of 3.75 to 1 to 0.52. These species become catalytically active during formic acid addition at 93 °C. HgO also reduces to elemental Hg once formic acid becomes available under acidic conditions. Catalytic activity was originally identified during lab-scale nuclear waste processing over fifteen years ago (1), and the potential for catalytic hydrogen generation was studied just prior to completion of the DWPF design. The chemical reaction catalyzed by the noble metals is given by:

$$HCOOH \rightarrow H_2 + CO_2$$

The planned DWPF air purge system, used to keep the process equipment under slight negative pressure, was redesigned to permit a large flow of purge air through the SRAT in order to dilute hydrogen below the lower flammable limit (LFL). Much subsequent experimental work was performed with the goal of ensuring that operating conditions would be selected that produced hydrogen concentrations below the LFL in the available air purge.

Catalytic hydrogen generation can be three or more orders of magnitude greater than radiolytic hydrogen generation when noble metals are at their most active state. This creates the potential for the formation of flammable or explosive air-hydrogen mixtures in the SRAT off-gas system. The DWPF analyzes the hydrogen concentration of the off-gas once per minute using a pair of redundant gas chromatographs. Catalytic hydrogen generation, however, can effectively be interrupted by dropping the temperature from 93-103 °C to less than 60 °C, whereas radiolytic hydrogen generation is not impacted by temperature.

Changes in the preparation of the legacy HLW sludges have been made to reduce the amount of decanted wash water. These less washed feeds have resulted in DWPF SRAT cycles that require proportionately more acid to process (more hydroxide, carbonate, nitrite per unit volume). This evolution in operating doctrine has constrained the quantity of acid that can be added during pre-treatment relative to the minimum required to accomplish various processing controls in the DWPF. This in turn has brought a renewed interest in understanding the nature of catalytic hydrogen generation and in seeking means to better control and/or eliminate it. Consequently, several new rounds of experimental work were authorized to better understand the process chemistry.

## EXPERIMENTAL PROGRAM

The need to produce a large volume of melter feed simulant that would require many lab-scale SRAT cycles provided an opportunity to design a matrix of tests the contained individual noble metals in the bulk sludge solids. These tests confirmed some earlier hypotheses about catalytic hydrogen generation. These tests, however, required the exclusion of mercury. Two subsequent tests were performed at smaller scale using a single noble metal combined with mercury in the bulk sludge solids. The results were used to design a statistical matrix of testing at smaller scale focused on the three most significant elemental species: Rh, Ru, and Hg. A full factorial design was conducted with each factor at either a high or low level (8 tests), plus three tests at the matrix midpoint. The same bulk sludge solids were available for all of these tests, in other words many factors that normally vary from test program to test program were held constant throughout these studies. The starting concentration of nitrite ion in the feed was the same during a series of over forty tests.

New data for the Ru-sludge system show that ruthenium does not become catalytically active for hydrogen generation in the presence of an excess of nitrite ion, but that ruthenium does become catalytically active once nitrite ion is brought to concentrations below that of ruthenium, Figure 1.

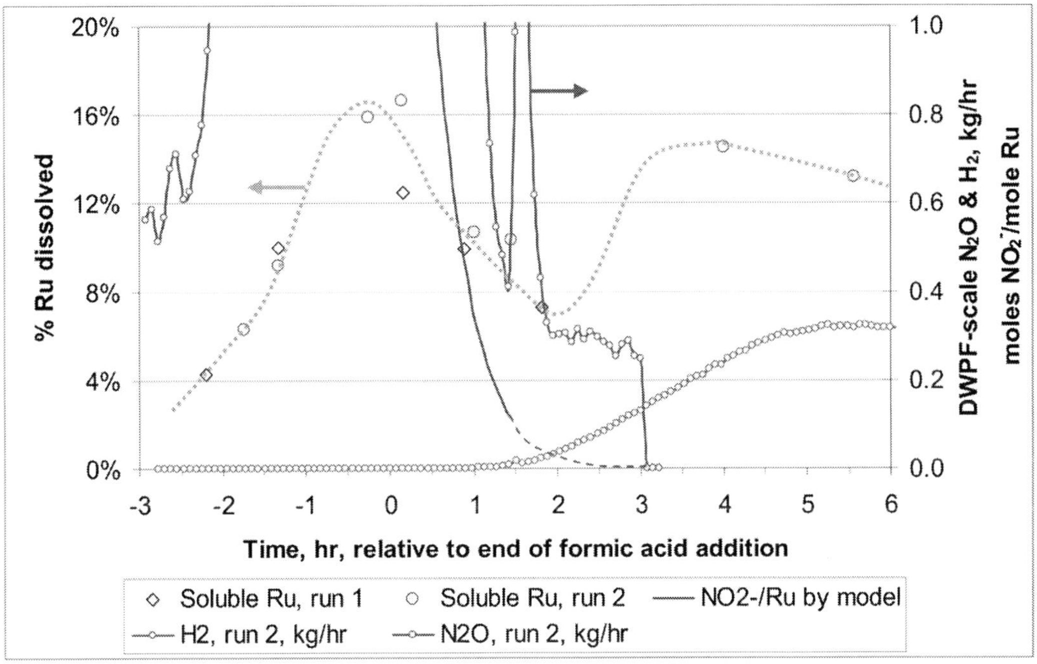


Figure 1. Relative timing of events related to Ru-catalyzed hydrogen generation in the absence of mercury, including nitrite destruction and associated $N_2O$ generation.

The nitrite ion concentration data correlate well with the $N_2O$ off-gas data with respect to the timing of complete nitrite destruction at about three hours after formic acid addition. The amount of nitrite ion present at the onset of hydrogen generation at one hour past formic acid addition is less than 0.5 mole nitrite ion per mole Ru, indicating that nitrite is losing its potential effectiveness as an inhibitor for Ru catalyzed hydrogen generation.

The addition of mercury to the single noble metal system with Ru produced several effects, Figure 2. The first was the accelerated conversion of nitrite to NO at the expense of $N_2O$ that was observed in the off-gas data. This shift was most obvious in the $N_2O$ data, but it was supported by additional evidence of $O_2$ depletion from the conversion of NO to $NO_2$. Second, nitrite destruction finished nearly an hour earlier in the Hg-Ru-sludge system than in the Ru-sludge system. As shown in Figure 2, the $N_2O$ peak ended about an hour earlier, and the onset of hydrogen generation coincided with the end of nitrite destruction in both cases.

The impact of an earlier onset of hydrogen generation had a minimal impact on the peak rate or the timing of the peak rate at about 5.5 hours after formic acid addition.

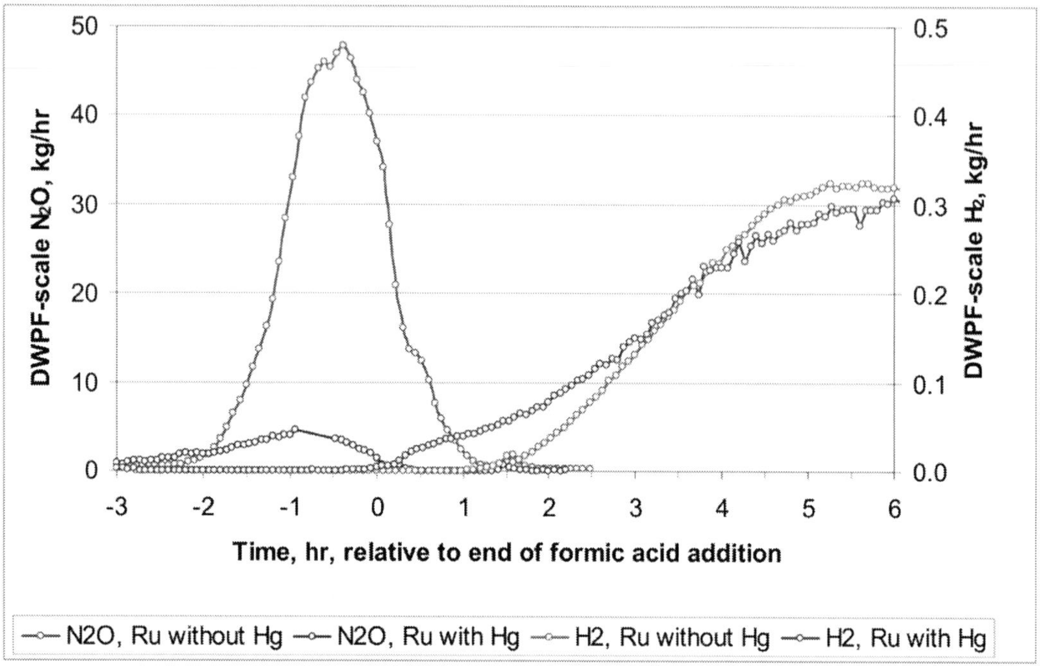


Figure 2. Impact of Hg on Ru catalysis of formic acid to produce hydrogen

Data on Ru in the supernate were obtained only after the onset of hydrogen in the Hg-Ru-sludge test, but the data overlaid the data from the two runs using Ru-sludge during this period. No evidence for a direct impact of Hg on the partitioning of the Ru between the supernate and insoluble solids was seen.

The most active catalytic form of Rh during the SRAT cycle appears to be a nitro-Rh complex. Such a complex was first proposed by King and Bhattacharyya during 1992-3 testing under contract to SRS and also in (2). Rh has now been identified as also catalyzing the destruction of nitrite ion to nitrous oxide and carbon dioxide during formic acid addition. This reaction probably involves one form of a nitro-Rh complex, $Rh(NO_2)_n^{3-n}$. The presence of Rh at 0.2 wt% in the total solids caused the production rate of $N_2O$ to more than double compared to tests with no noble metals. At high nitrite ion concentrations the Rh is believed to be present as a hexanitro-Rh complex, $Rh(NO_2)_6^{3-}$. Gradual destruction of nitrite ion leads to eventual substitution of one or more of the nitrite ligands with other species ($H_2O$, $OH^-$, formate, etc.).

New data show that it is during the period near the end of nitrite destruction that Rh becomes an excellent catalyst for converting formic acid into hydrogen and carbon dioxide. Figure 3 shows, however, that the Rh catalyzed hydrogen generation period occurs before the nitrite is entirely destroyed.

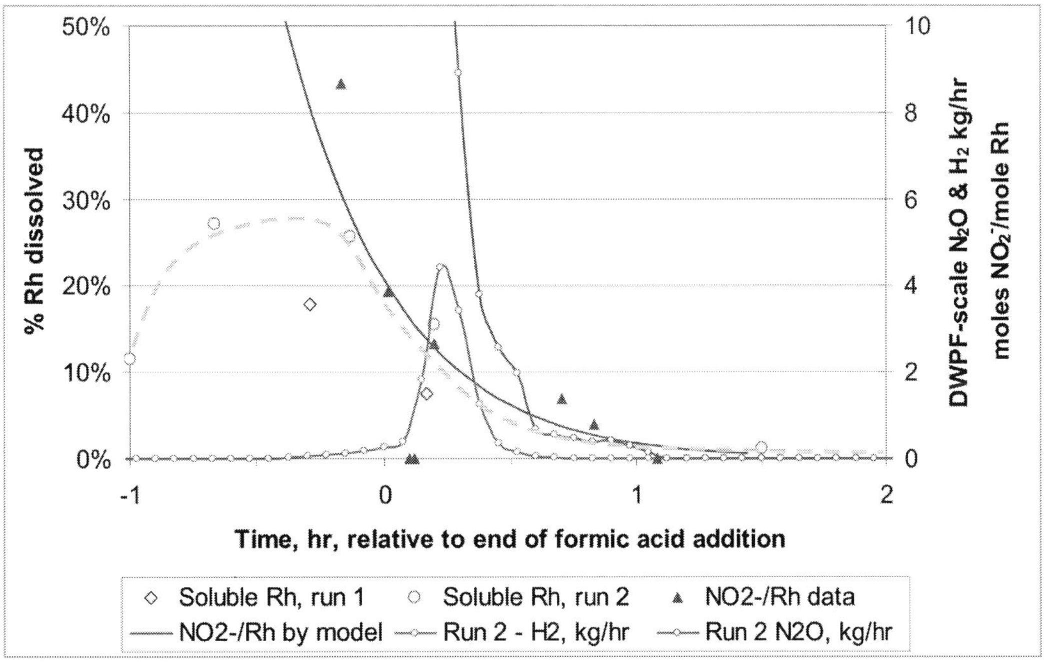


Figure 3. Hydrogen generation goes through a maximum near the time when the ratio of total nitrite to total Rh is between two and four.

The Rh is partitioned between the aqueous and insoluble solid phases, but it is not clear whether or not the nitro-Rh complex is stable and active adsorbed onto the insoluble solid surfaces or is only active in the aqueous phase. The ratio of nitrite to Rh in the aqueous phase alone could be as high as twenty to one at the maximum hydrogen generation rate based on the available data. In any case, a condition is reached where one or more of the ligands can be substituted with formate and/or other ions to permit the conversion of formic acid to hydrogen. One unresolved analytical issue is whether or not the nitrite ions complexed onto the Rh can be detected by ion chromatography (the nitrite ion has to be solubilized for detection). $N_2O$ data are included in Figure 3 to show that nitrite destruction was not completed prior to the hydrogen generation rate maximum at about ten minutes after formic acid addition.

The virtual completion of nitrite destruction (by the $N_2O$ trace) about 30-35 minutes after formic acid addition corresponds with the end of the peak in the hydrogen generation rate. This indicates that the initial deactivation mechanism for the nitro-Rh complex is the elimination of the nitrite ligands and simultaneous destruction of the nitro-Rh complex. The kinetics of nitrite destruction appears to be controlling both the activation and the deactivation of the nitro-Rh complex form that catalyzes hydrogen generation. The nitro-Rh complex is currently the catalyst of primary concern. Although there is more Ru than Rh in HLW, the activity of the nitro-Rh species exceeds that of whatever Ru species catalyzes hydrogen generation (suspected to be either a Ru-carbonyl complex or a hydrolyzed Ru complex).

Adding Hg to the Rh-sludge system caused some adjustments to the nitrite destruction and hydrogen generation as it did in the Ru-sludge system, Figure 4.

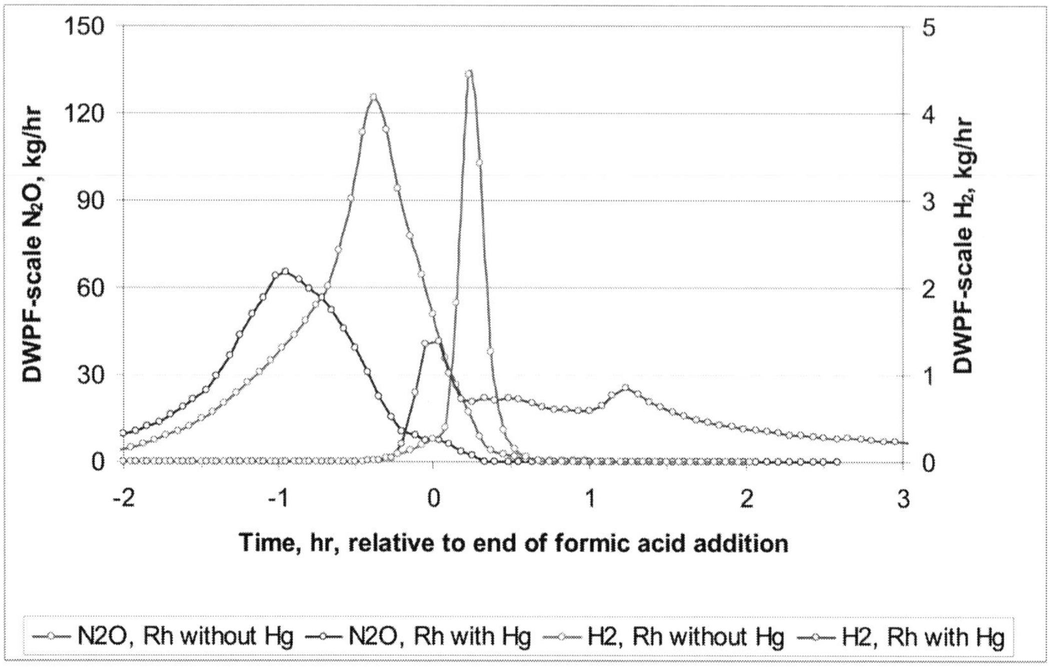


Figure 4. Impact of Hg on the Rh-sludge system including accelerated nitrite destruction and earlier hydrogen generation.

Nitrite destruction was accelerated by the addition of HgO to the starting sludge simulant. A reduction in $N_2O$ formation was seen, but not as much as in the Ru-sludge case. Other testing has shown the Rh is a catalyst for $N_2O$ production. In the Hg-Rh-sludge system, Rh and Hg were competing to catalytically convert nitrite to either $N_2O$ or NO, while in the Hg-Ru-sludge system Hg was the only catalyst appreciably active during nitrite destruction.

Accelerated nitrite destruction in the Hg-Rh-sludge system caused initiation of the Rh catalyzed hydrogen generation reaction to occur sooner. In this case, the initiation came before all of the formic acid had been added. In other words, less of the reactant was present when Rh activated for hydrogen generation. This may explain why the maximum hydrogen generation rate obtained with Hg present was only a third that obtained when Hg was absent. The Hg-Rh-sludge system retained significant catalytic activity for hydrogen generation after nitrite destruction. The Rh-sludge system destroyed enough formic acid that the pH rose above seven about four hours after formic acid addition, but the Hg-Rh-sludge system retained enough formic acid that it stayed below pH six for fifteen hours after formic acid addition. The sustained catalytic activity in the Hg-Rh-sludge system may be due to the presence of much more formic acid relative to sodium formate.

The presence of nitrite ion during the period of maximum Rh activity coupled with the absence of nitrite ion during the period of Ru activity for hydrogen generation indicate that the two catalysts are unlikely to combine to produce an enhanced maximum generation rate. The slower activation of Ru for catalytic hydrogen generation often correlates with lower hydrogen generation rates than occurred during the period of maximum Rh activity, but this is not always true.

Rh retains some activity after destruction of the nitro-Rh complex. A number of lab-scale process simulations have shown continued lower levels of hydrogen generation after complete nitrite destruction. A primary deactivation mechanism for either Ru catalysis or the post-complex Rh species has not been fully identified, but adsorption and/or precipitation of both soluble Rh and soluble Ru onto the insoluble solids correlates linearly with reductions in the observed catalytic hydrogen generation rate. Falling hydrogen generation rates in both cases are observed to coincide with decreasing concentrations of the soluble form(s), but the soluble forms are typically only 1-5% of the total Rh or Ru during this period.

Palladium has been found to be relatively ineffective at converting formic acid into hydrogen in SRAT simulations, which, coupled with its lower concentrations, makes it a minor factor for hydrogen generation. Tests of Pd at 0.2 wt% compared to Rh at 0.2 wt% in the total solids produced maximum hydrogen generation rates that differed by over a factor of 1000. Pd has been found, however, to be a fairly effective catalyst for formic acid conversion of nitrite ion to nitrous oxide, carbon dioxide, and water, in other words Pd acts like Rh during nitrite destruction and with comparable catalytic activity. Therefore, Pd influences the Rh and Ru catalysis of formic acid to hydrogen via its impact on the nitrite ion.

The new data on the individual roles of Rh, Ru, Pd, Hg and nitrite ion presented in this talk represent a considerable advance in the understanding of the reaction sequences and the interplay of the significant species during processing of the legacy waste slurry upstream of the DWPF melter. Testing on systems with all species present simultaneously exhibit all of the above effects. The new data have identified the causes for the effects that were previously observed but only poorly understood.

REFERENCES

1. Noble Metal Catalyzed Hydrogen Generation from Formic Acid in Nitrite-Containing Simulated Nuclear Waste Media, R. B. King, N. K. Bhattacharyya, and K. D. Wiemers, Environmental Science and Technology, Vol. 30, No. 4, (1996) 1292-1299.

2. Catalytic reactions of formate 4. A nitrite-promoted rhodium(III) catalyst for hydrogen generation from formic acid in aqueous solution, R. B. King and N. K. Bhattacharyya, Inorganica Chimica Acta, 237 (1995) 65-69.

3. Koopman, D., B. Pickenheim, D. Lambert, J. Newell, and M. Stone, "Characterization of Individual Chemical Reactions Consuming Acid during Nuclear Waste Processing at the Savannah River Site", Proceedings of the 2009 Annual AIChE Meeting, New York, NY: AIChE. 2009.

# Wet Air Oxidation Batch Autoclave Testing with Savannah River Site Tank 48H Actual Waste

Kofi Adu-Wusu, Environmental & Chemical Process Technology, Savannah River National Laboratory, Aiken, SC

Paul R. Burket, Environmental & Chemical Process Technolgy, Savannah River National Laboratory, Aiken, SC

Chris J. Martino, Environmental & Chemical Process Technolgy, Savannah River National Laboratory, Aiken, SC

Frank M. Pennebaker, Environmental & Chemical Process Technolgy, Savannah River National Laboratory, Aiken, SC

Tank 48H at the Department of Energy's (DOE) Savannah River Site (SRS) in Aiken, South Carolina currently holds legacy material containing organic tetraphenylborate (TPB) compounds from the operation of the In-Tank Precipitation process. This material is not compatible with the waste treatment processes at SRS and must be removed or undergo treatment to destroy the organic compounds before the tank can be returned to Tank Farm service. In addition, the material can decompose to benzene and lead to potentially flammable concentrations in the tank headspace. Tank 48H currently holds approximately 250,000 gallons of alkaline slurry containing 21,800 kg of potassium and cesium tetraphenylborate (KTPB and CsTPB) solids. The tank has been isolated from Tank Farm service, and its return to service is a high priority to the DOE.

Wet Air Oxidation (WAO) is one of the two technologies currently under consideration for the treatment of TPB in Tank 48H. WAO is an aqueous phase process in which soluble and/or suspended waste components are oxidized using oxygen or oxygen in air. In general, the process operates at elevated temperatures and pressures ranging from about 150 to 320 °C and 7 to 210 atmospheres, respectively. The products of the reaction are carbon dioxide, water, and low molecular weight oxygenated organics (e.g. acetate, oxalate, etc.)

Batch bench-scale autoclave testing with actual (radioactive) Tank 48H waste is among the tests required in the DOE WAO Technology Maturation Plan. The objective of the testing is to confirm the ability of the WAO process to destroy TPB and its associated compounds using actual (radioactive) Tank 48H waste. Specifically: (i) Confirm the TPB destruction efficiency obtained from comparable simulant tests at same reaction times, (ii) Determine the destruction efficiency of other organics including biphenyl, (iii) Identify and quantify the reaction byproducts, and (iv) Determine off-gas composition.

Batch bench-scale stirred autoclave tests using both simulated and actual Tank 48H wastes have been conducted. The study was approached in three ways. (1) Replication of the 2006 batch bench-scale shaking autoclave simulant tests at the same conditions as those identified in 2006; (2) Batch bench-scale stirred autoclave simulant tests at nearly the same (i.e, conservative) conditions identified during the continuous-flow pilot-scale testing conducted in the January 2009; and (3) Replication of item # 2 above using actual or radioactive Tank 48H waste.

The test conditions for the first approach were a temperature of 300 °C, a reaction time of 3 hours, $CuSO4.5H2O$ solution (500 mg/L Cu) as catalyst, and 1:1 Tank 48H simulant dilution with 2M NaOH/$CuSO4.5H2O$ solutions, and 1 mL/L of antifoam agent.

The test conditions for the second and third approach were a temperature of 280 °C, a reaction time of 1 hour, $CuSO4.5H2O$ solution (250 mg/L Cu) as catalyst, and 1:1 Tank 48H material (simulant or radioactive waste) dilution with 2M NaOH/$CuSO4.5H2O$ solutions, and 0.21 mL/L of antifoam agent.

The test conditions for the second and third approach are conservative because the continuous-flow pilot-scale testing used a dilution ratio of 2:1. Note that even though a catalyst concentration of 100 mg/L Cu was used in the pilot-scale testing, the 250 mg/L Cu used in approaches 2 and 3 testing is not expected to impact the performance. Only the soluble form of the Cu performs the catalytic action. The amount of Cu that remains soluble is less than 100 mg/L.

The tests of the first approach demonstrated a good replication of the results of the 2006 testing. TPB, its daughter compounds (triphenylborane [3PB], diphenylborinic acid [2PB], phenylboronic acid [1PB]), phenol, and biphenyl, in the simulant were all destroyed to below their respective detection limits. It must be noted that biphenyl was only partially destroyed in the 2006 shaking autoclave testing. The almost total destruction of biphenyl is attributed to the better mixing characteristic of stirred autoclaves. Other results in terms of conversion of nitrite to nitrate, solid compounds in the treated simulant identified by x-ray diffraction, constituents in the off-gas, etc. were largely comparable.

The tests of the second and third approach demonstrated WAO is a viable technology for the treatment Tank 48H radioactive waste. TPB, 3PB, 2PB, 1PB, phenol, nitrobenzene, nitrosobenzene, etc. in the radioactive waste were all destroyed to below their respective detection limits. In fact, biphenyl is the only organic compound of interest that was not completely destroyed. The partial destruction of biphenyl stems from its propensity to go to the vapor phase during the oxidation reaction.

The undestroyed biphenyl is essentially biphenyl vapor that solidifies (because of its low melting point - about 68 oC) when the autoclave is cooled at the end of the WAO reaction. The undestroyed biphenyl is expected to be in the off-gas in the actual WAO process, which is a continuous-flow system.

Destruction efficiency of TPB was > 99.997% while that of the other organic compounds (except biphenyl) were for the most part > 85%. The fairly low destruction efficiencies (compared to that of TPB) for most of the other compounds are due to their relatively high detection limits and/or low initial concentrations in the waste.

The autoclave radioactive waste tests were a confirmation to earlier simulant tests performed at identical conditions. The radioactive waste tests and the simulant tests were fairly comparable particularly, the destruction efficiencies of the major organic compounds (TPB and phenol).

All the above results should be viewed in the context that the test conditions are conservative compared to the conditions identified in the continuous-flow pilot-scale testing. By all accounts, had the tests been conducted at a 2:1 diluent/Tank 48H waste volume ratio as identified in the pilot-scale testing (versus the 1:1 diluent/Tank 48H waste volume ratio used for the current tests), the destruction efficiency of all the organic constituents in the Tank 48H waste especially biphenyl most likely would have been greater.

The common compounds identified by x-ray diffraction (XRD) in all four tests (i.e., two radioactive waste tests and two simulant tests) were CuO (tenorite), carbon (graphite), and Na2Ti3O7 (sodium titanium oxide). Graphite is an artifact of the tests. It is the result of autoclave graphite gasket breakage and possibly erosion from graphite stirrer bushing.

Additional compounds identified by XRD in radioactive tests 1 and 2 were H2(Ti2SiO7)(H2O)1.5 (hydrogen titanium silicate hydrate) and FeO(OH) (goethite) respectively. Even though these compounds were not identified in the simulant tests, they are all believed to present in all the treated material. Smaller loss of solids during pouring of treated material in the simulant tests compared to the radioactive tests precluded their identification by XRD.

The percent conversion of nitrite to nitrate was higher in the simulant tests (17%) than radioactive tests (5%). The organics-oxygen reaction from the increased number of organic constituents or compounds (oxalate, formate, nitrobenzene, etc.) in the radioactive waste seems to supercede the nitrite-oxygen reaction.

The major radionuclide in the waste, Cs-137, is present in the treated waste as soluble species even though it is in the untreated waste as 137CsB(C6H5)4 solids. The same is true for B and K which are also present in the untreated waste as insoluble 137CsB(C6H5)4 and KB(C6H5)4 solids. Cs-133, the non radionuclide cesium in the waste behaves in a similar manner. Np-237, U-233, U-233, U-234, U-235, U-236, U-238, and Pu-238 are largely present in the treated waste as soluble species while Sr-90 is present in insoluble form.

The untreated off-gas from both the radioactive waste and the simulant tests indicated the major constituents are nitrogen, oxygen, benzene, and hydrogen in that order. All other gases (methane/other volatile organic compounds, carbon dioxide, carbon monoxide, nitric oxide, and nitrious oxide) were either below detection limits or were not detected at all. Benzene and hydrogen in the off-gas for all the tests were relatively low. The highest concentrations of benzene and hydrogen in the off-gas obtained for all four tests were 0.096 and 0.0063 vol% respectively. These translate to 6.9% of benzene lower flammability limit (LFL) of 1.4 vol% at 25 oC and 0.16% of hydrogen LFL of 4 vol% at 25 oC.

See more of this Session: Chemical Engineering Advances in the Processing of Radioactive Waste
See more of this Group/Topical: Nuclear Engineering Division

# 2009 THOR® (Thermal Organic Reduction) Fluidized Bed Steam Reforming Engineering Scale Test

**Gene Daniel**, Engineering Process Development, Savannah River National Laboratory, Aiken, SC

C. M. Jantzen, Pro. Tech. Prgm., Savannah River National Laboratory, Aiken, SC

M. R. Williams, Eng. Dev. Lab, Savannah River National Laboratory, Aiken, SC

The Savannah River National Laboratory (SRNL) will perform lab analyses on samples from the June 2009 THOR® Fluidized Bed Steam Reforming (FBSR) Engineering Scale Test Demonstration (ESTD) Phase 3 with Tank 48H simulant at the Hazen Research Inc. (HRI) in Golden, Colorado. This work is in support of the Savannah River Remediation Liquid Waste Operations for Tank 48 at the Savannah River Site. The purpose of the 2009 ESTD Phase 3 testing at the Hazen facility is to demonstrate the long term operability of an integrated FBSR processing system with product solids handling capability. The THOR® Treatment Technologies (TTT) FBSR process of destroying organics in the Tank 48 waste simulant and generating a solid carbonate product was successfully demonstrated on the pilot scale level in 2006. Pilot scale tests were completed in 2008 (Phase 1) and in 2009 (Phase 2) that further demonstrated the TTT FBSR process ability to destroy organics in the Tank 48 simulant and produce a transportable solid carbonate waste form. The final Phase 3 pilot scale testing will demonstrate the integrated TTT FBSR process capability to process the Tank 48 simulant from a slurry feed into a greater than 99% organic free and primarily dissolved carbonate FBSR product slurry. The Phase 3 analytical results will be compared to prior 2008 and 2006 pilot scale results in terms of organic, nitrite, and nitrate destruction.

See more of this Session: Chemical Engineering Advances in the Processing of Radioactive Waste
See more of this Group/Topical: Nuclear Engineering Division

# Removing Sludge Heels From SRS Waste Tanks with Oxalic Acid

Michael Poirier, Savannah River National Laboratory, Aiken, SC

Presentation will decribe removing sludge heels from SRS waste tanks.

See more of this Session: Chemical Engineering Advances in the Processing of Radioactive Waste
See more of this Group/Topical: Nuclear Engineering Division

# Boehmite Dissolution Model Based On Simulant Data

**Reid A. Peterson**, Pacific Northwest National Laboratory, Richland, WA

Renee Russell, Pacific Northwest National Laboratory, Richland, WA

Boehmite is present in significant quantities in several of the Hanford waste tanks. It has been proposed that the boehmite will be dissolved through caustic leaching in the Hanford Waste Treatment and Immobilization Plant (WTP) currently under construction. Therefore, it is important to fully understand the nature of this dissolution process so that caustic leaching can be effectively deployed on the Hanford tank wastes.

This research determined the impact of primary control parameters such as temperature, hydroxide concentration, approach to solubility, and stirring rate on the boehmite dissolution rate. The impact of aluminate ion on the dissolution kinetics was also determined. In addition, other parameters that impact boehmite dissolution, such as free-hydroxide concentration and reaction temperature, were also assessed and used to develop a semi-empirical model of the boehmite dissolution process. The understanding derived from this work will be used as the basis to evaluate and improve the planned performance of aluminum leaching during waste pretreatment in preparation of waste vitrification in the WTP .

This work is the first in a series of programs aimed at demonstrating the WTP waste pretreatment leaching process . This work was used to develop a simulant of the boehmite-containing Hanford waste. That simulant is subsequently being used in laboratory- and pilot-scale testing to demonstrate the WTP pretreatment process in an integrated fashion.

See more of this Session: Chemical Engineering Advances in the Processing of Radioactive Waste
See more of this Group/Topical: Nuclear Engineering Division

## Solidification Tests for LLW Sludges

*Paul A. Taylor, Oak Ridge National Laboratory, Oak Ridge, TN, USA*

### Background

The Oak Ridge Transuranic Waste Processing Center (TWPC), operated by EnergX TN, LLC, must process about 350,000 gal of remote-handled (RH) sludge from ten liquid low-level waste (LLLW) tanks at Oak Ridge National Laboratory (ORNL). In order to meet processing requirements and the Waste Acceptance Criteria for the Nevada Test Site (NTS), the waste must be mixed with solidification/stabilization agents, remain flowable during mixing, be self-leveling in the waste disposal container, and produce a solid waste form that is not a hazardous waste and has no free liquids, suitable for transportation and disposal at NTS. Laboratory-scale grouting tests were performed on samples of sludge from four tanks, to evaluate the proposed grouting formulation.

### Materials and Methods

Core samples of the waste sludge were collected from four tanks at ORNL (W-23, W-25, W-28, and W-35), and placed in a hot cell. The tubes from each tank (2-3 per tank) were combined and mixed to give four homogenous sludge samples. Ten sub-samples were collected from each blended sample, one for physical, chemical, and radionuclide analysis and the rest for grout testing. The grouting tests were conducted inside a glove box, using an electric mixer to blend the samples. The mixer and impeller are shown in Fig. 1. The impeller was rotated clockwise (from above) to pump the slurry down through the impeller and up along the sides of the bottle.

The sludge samples were diluted with water to a total suspended solids (TSS) concentration of 5 wt% before grouting, which is the planned concentration for sluicing the sludge from the tanks for treatment. Figures 2 and 3 show concentrated and diluted sludge samples, respectively. One sample from each tank was grouted at a TSS concentration of 10 wt%, as a worst-case test of hazardous metals stabilization. The dry-blend solidification ingredients included Portland cement (Type I/II, Cemex Corp., Knoxville, TN), fly ash (Type F, Cumberland Steam Plant, Tennessee Valley Authority), ground blast furnace slag (GBFS, Grade 120, LaFarge, Corp., Grand Chain, IL), and small amounts of two proprietary additives.

The procedure for grouting the samples was as follows.

1. Weigh sludge sample (~10 g) and calculate the amount of water needed to produce the desired TSS concentration. Use the water to rinse the sludge from the sample bottle into a 250-mL plastic bottle, and mix the sludge and water

2. Calculate the amount of cement, fly ash, and GBFS needed, weigh out the required amounts, and add to the diluted sludge slurry
3. Stir for 5 minutes at 500 rpm
4. Calculate the amount of additive #1 needed, weigh, add to the slurry, and stir for 5 minutes at 500 rpm
5. Calculate the amount of additive #2 needed, weigh, add to the slurry, and stir for 5 minutes at 500 rpm
6. Stop stirrer, add ~5 mL of grout mixture to each of two small plastic bags, and flatten the grout in the bags, for analytical samples
7. Monitor the grout in the bottle for free water and solidification

Fig. 1. Mixer and impeller

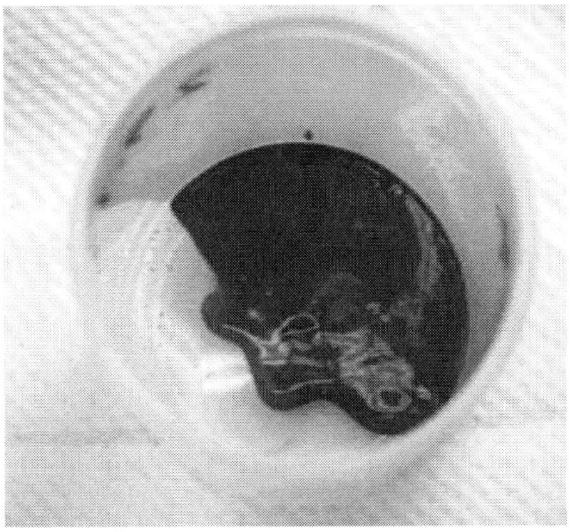

Fig. 2. Concentrated sludge (19.8 wt% TSS)

Fig. 3. Diluted sludge (5.0 wt% TSS)

Figure 4 shows a grouted sample as it is starting to solidify, and Fig. 5 shows the subsamples in the plastic bags for TCLP analysis.

Fig 4. Grouted sample

Fig 5. Subsamples in plastic bags

After curing, one of the samples in the plastic bags from each test was transferred to the ORNL Radioactive Materials Analysis Laboratory (RMAL), where a portion of the grout was broken up by hand to meet the size requirements (<9.5 mm) of the Toxicity Characteristics Leaching Procedure (TCLP). The size-reduced grout sample was extracted, and the extractant was analyzed for cadmium, chromium, and lead by inductively-coupled plasma-mass spectroscopy (ICP-MS, EPA Method 200.8) and for mercury by cold vapor atomic adsorption (CVAA, EPA Method 245.1). The TCLP procedure results in a dilution of 20× for contaminants in the extraction solution compared to the concentration in the starting solid matrix. Any reduction in concentration beyond 20× would be due to the matrix retaining the contaminants during the extraction step.

## Results and Discussion

The analytical results for the blended sludge samples from the four waste tanks are shown in Table 1. Adding water to the concentrated sludge, to reduce the TSS concentration prior to grouting, dissolves a portion of the existing solids, as a new equilibrium is established with the more dilute liquid. To determine the actual amount of water that will be needed to produce a 5 or 10 wt% TSS slurry, RMAL took a small portion of the blended sludge from each tank, added water to produce a dilute slurry, and then measured the actual TSS concentration in the slurry. Comparing the measured TSS concentration to the predicted concentration in the diluted slurry, calculated from the TSS concentration in the concentrated sludge (see Table 2) and the amount of water added, gives an "effective TSS" in the concentrated sludge samples.

**Table 1.** Summary of analytical results for sludge samples

| Analyte | Tank W-23 | Tank W-25 | Tank W-28 | Tank W-35 |
|---|---|---|---|---|
| Density (g/mL) | 1.267 | 1.284 | 1.227 | 1.190 |
| TSS[a] (wt%) | 31.5 | 38.7 | 24.0 | 19.7 |
| pH (units) | 13.23 | 12.11 | 11.84 | 9.85 |
| | | | | |
| Cr (µg/g) | 241 | 179 | 112 | 199 |
| Cd (µg/g) | 4.4 | 15.6 | 26.0 | 6.2 |
| Pb (µg/g) | 205 | 427 | 333 | 447 |
| Hg (µg/g) | 29.0 | 85.9 | 44.1 | 49.9 |
| | | | | |
| Am-241 (Bq/g) | 3.80E+03 | 6.70E+03 | 6.10E+03 | 5.70E+03 |
| Cm-244 (Bq/g) | 1.41E+05 | 8.20E+03 | 4.41E+04 | 3.32E+04 |
| Cs-137 (Bq/g) | 1.70E+06 | 3.40E+05 | 2.00E+05 | 1.00E+06 |
| Eu-154 (Bq/g) | 1.40E+04 | 6.10E+04 | 9.70E+04 | 4.00E+04 |
| Eu-155 (Bq/g) | ND[b] | 6.40E+03 | 7.70E+03 | ND |
| Pu-239 (Bq/g) | ND | ND | ND | 4.18E+03 |
| Pu-238 (Bq/g) | 1.49E+04 | 7.10E+04 | 4.79E+03 | ND |
| Sr-90 (Bq/g) | 6.90E+05 | 1.90E+06 | 1.10E+06 | 1.20E+06 |
| | | | | |
| U-238 (µg/g) | 2.73E+04 | 1.83E+04 | 2.21E+04 | 2.77E+04 |
| Pu-239 (µg/g) | 1.73E+00 | 2.18E+00 | 1.64E+00 | 3.03E+00 |
| Pu-240 (µg/g) | 2.16E-01 | 3.28E-01 | 2.61E-01 | 2.51E-01 |
| Am/Pu-241(µg/g) | 4.50E-02 | 8.42E-02 | 8.31E-02 | 6.00E-02 |
| Am/Pu-242 (µg/g) | 9.60E-02 | 4.91E-02 | 3.54E-02 | 3.84E-02 |

[a]TSS = Total Suspended Solids (EPA method 600-160.1)

[b]Not Detected

The equation for the predicted TSS concentration in the diluted slurries is as follows:

$$\textit{Predicted Dilute TSS (wt\%)} = \frac{\textit{Sludge Weight (g) X TSS(wt\%)}}{\textit{Sludge Weight (g)} + \textit{Dilution Water Weight (g)}}$$

The equation for calculating the effective TSS concentration is as follows:

$$Effective\ TSS\ (wt\%) = \frac{Sludge\ TSS\ (wt\%)\ X\ Measured\ dilute\ TSS\ (wt\%)}{Predicted\ dilute\ TSS\ (wt\%)}$$

The effective TSS value was used to calculate the amount of water needed to produce the diluted slurries for the grouting tests. Table 2 shows the results of these measurements and calculations. The effective TSS concentration is lower than the measured TSS by 21–47%. The equation for calculating the amount of water to add to a sludge sample to obtain a desired TSS concentration in the dilute slurry is as follows:

$$Water\ needed\ (g) = \frac{Sludge\ weight\ (g)\ X\ Effective\ TSS\ (wt\%)}{Desired\ TSS\ (wt\%)} - Sludge\ weight\ (g)$$

**Table 2.** Results of slurry TSS measurements and calculations

| Slurry property | W-23 | W-25 | W-28 | W-35 |
|---|---|---|---|---|
| Sludge TSS | 31.5 | 38.7 | 24.0 | 19.7 |
| Sludge weight (g) | 10.023 | 10.033 | 10.083 | 10.013 |
| Water added (g) | 63.03 | 77.45 | 47.92 | 39.49 |
| Predicted TSS | 4.32 | 4.43 | 4.14 | 3.99 |
| Measured TSS | 2.54 | 2.34 | 3.27 | 2.96 |
| Effective TSS | 18.55 | 20.45 | 18.96 | 14.67 |

The grouting formulation used for these tests, with ratios of grout-forming additives weight to diluted slurry weight ranging from 0.75:1 to 1.2:1, produced wet grout mixtures that were easy to stir and were self-leveling. The grout mixtures cured with no free water visible at any time. The cured grout matrix was very effective at retaining the hazardous metals in the sludge, with TCLP leachate concentrations well below the limits in the Resource Conservation and Recovery Act (RCRA) (see Table 3). The weight fraction of concentrated sludge in each grout mixture is also shown in Table 3, which is dependent on the diluted sludge TSS concentration and the ratio of grout-forming additives to diluted sludge for each test. The TCLP extraction was performed on three samples after 1 day and 7 days of curing, compared to the >28 days for the remaining samples. The metal concentrations for the short cure time samples were similar to the remaining samples and were all well below the RCRA limits, which are 5.0 mg/L for Cr and Pb, 1.0 mg/L for Cd, and 0.2 mg/L for Hg.

**Table 3.** Analytical results for TCLP metals in grout samples

| Sample ID | Sludge fraction | Metals concentration (mg/L) | | | |
|---|---|---|---|---|---|
| | | Cd | Cr | Pb | Hg |
| W25-1 | 0.064 | <0.03 | 0.558 | <0.04 | 0.0084 |
| W25-2 | 0.068 | <0.03 | 0.682 | <0.04 | 0.0127 |
| W25-3 | 0.122 | <0.03 | 0.629 | <0.04 | 0.0284 |
| W25-4 | 0.162 | <0.03 | 0.584 | <0.04 | 0.0289 |
| W25-5 | 0.111 | <0.03 | 0.570 | <0.04 | 0.0165 |
| W25-6 | 0.122 | <0.03 | 0.529 | <0.04 | 0.0176 |
| W25-7 | 0.122 | <0.03 | 0.567 | <0.04 | 0.0231 |
| W25-8 | 0.244 | <0.03 | 0.516 | 0.0922 | 0.0352 |
| W25-9 | 0.122 | <0.03 | 0.630 | <0.04 | 0.0193 |
| | | | | | |
| W28-1 | 0.131 | <0.03 | 0.422 | <0.04 | 0.0083 |
| W28-2 | 0.149 | <0.03 | 0.573 | <0.04 | 0.0103 |
| W28-3 | 0.119 | <0.03 | 0.393 | <0.04 | 0.0072 |
| W28-4 | 0.131 | <0.03 | 0.699 | 0.0709 | 0.0078 |
| W28-5 | 0.130 | <0.03 | 0.437 | 0.0976 | 0.0053 |
| W28-6 | 0.130 | <0.03 | 0.316 | <0.04 | 0.0056 |
| W28-7 | 0.262 | <0.03 | 0.507 | <0.04 | 0.0106 |
| W28-8 | 0.132 | <0.03 | 0.639 | 0.0634 | 0.0063 |
| | | | | | |
| W35-1 | 0.169 | <0.03 | 0.790 | <0.04 | <0.005 |
| W35-2 | 0.193 | <0.03 | 1.140 | <0.04 | 0.0078 |
| W35-3 | 0.153 | <0.03 | 0.557 | <0.04 | <0.005 |
| W35-4 | 0.169 | <0.03 | 0.615 | <0.04 | <0.005 |
| W35-5 | 0.169 | <0.03 | 0.643 | <0.04 | <0.005 |
| W35-6 | 0.338 | <0.03 | 0.956 | <0.04 | 0.0061 |
| W35-7 | 0.169 | <0.03 | 0.634 | <0.04 | <0.005 |
| W35-8 | 0.169 | <0.03 | 0.563 | <0.04 | 0.0054 |
| | | | | | |
| W23-1 | 0.133 | <0.03 | 0.721 | <0.04 | <0.005 |
| W23-2 | 0.152 | <0.03 | 0.745 | <0.04 | <0.005 |
| W23-3 | 0.121 | <0.03 | 0.462 | <0.04 | <0.005 |
| W23-4 | 0.132 | <0.03 | 0.596 | <0.04 | <0.005 |
| W23-5 | 0.133 | <0.03 | 0.601 | <0.04 | <0.005 |
| W23-6 | 0.266 | <0.03 | 0.386 | <0.04 | <0.005 |
| W23-7 | 0.135 | <0.03 | 0.734 | <0.04 | <0.005 |

| | | | | | |
|---|---|---|---|---|---|
| W23-8 | 0.134 | <0.03 | 0.532 | 0.171 | <0.005 |

**Table 4.** Analytical results for TCLP metals in grout samples after short cure times

| Sample ID | Cure Time (d) | Metals concentration (mg/L) | | | |
|---|---|---|---|---|---|
| | | Cd | Cr | Pb | Hg |
| W-28-9 | 1 | <0.03 | 0.593 | <0.04 | 0.0053 |
| W-28-9 | 7 | <0.03 | 0.549 | <0.04 | <0.005 |
| | | | | | |
| W-35-9 | 1 | <0.03 | 0.893 | <0.04 | <0.005 |
| W-35-9 | 7 | <0.03 | 0.814 | <0.04 | <0.005 |
| | | | | | |
| W-23-9 | 1 | <0.03 | 0.955 | <0.04 | <0.005 |
| W-23-9 | 7 | <0.03 | 0.747 | <0.04 | <0.005 |

As was the case with the previous samples with long cure times, there was no detectable cadmium in any of these samples. There was also no detectable lead, as was the case for most of the earlier samples. Chromium was detected in all of the samples, and the concentrations decreased slightly as the cure time increased, particularly for the sample from Tank W-23.

The results of these tests show that the grout formulation meets all of the requirements for solidifying the sludge from the LLW tanks at ORNL.

# History of Radioactive Waste - How Did We Get Here?

Ronald A. Palmer, Institute for Clean Energy Technology, Mississippi State University, Starkville, MS

Radioactivity was discovered by Henri Becquerel in 1896. Marie and Pierre Curie subsequently investigated the phenomenon and discovered polonium and radium in 1898. Over a century later now, we may well ask "What did they do with their garbage?" From those early days to Hanford, Savannah River, and West Valley today, the management of radioactive materials has followed a tortuous path. Although you can be sure that the early scientists did not want to hurt themselves, apparently not much thought went into the potentially hazardous effects of new materials or experiments. This paper will present a brief overview of the history of radioactive waste management.

See more of this Session: Chemical Engineering Advances in the Nuclear Fuel Cycle (including NED student competition)
See more of this Group/Topical: Nuclear Engineering Division

# Nuclear Chemical Engineering Today

**Stuart Arm**, Energy Solutions, Richland, WA

This presentation will show where and how nuclear chemical engineers are playing an important role in restoring the environment and meeting the nation's energy needs. Reference will be made to the Hanford and Savannah River DOE sites and the recycling of spent nuclear fuel.

See more of this Session: Chemical Engineering Advances in the Nuclear Fuel Cycle (including NED student competition)
See more of this Group/Topical: Nuclear Engineering Division

# The Risk of Nuclear Power: Weighing the Safety and Proliferation Concerns Associated with Nuclear Energy against the Threat of a Changing Climate and Energy Security

Sarah Widder, Chemical Engineering, University of Washington, Seattle, WA

In 2007, the United States consumed 42.09 quadrillion BTUs of electrical energy and produced 6 billion metric tons of carbon dioxide. The global warming impact of carbon dioxide has lead to a recent government initiative to require emissions reductions of 20% below 2005 levels by 2020. The United States will need "carbon-free" energy generation technologies that are available on a realistic timescale, and for economically competitive prices, to achieve this CO2 reduction.

Nuclear energy is largely "carbon free," and is the second most abundant electrical energy resource in the world. Tripling nuclear power energy generation capacity to 1 million megawatts could decrease carbon dioxide emissions by 0.8 to 1.8 billion tons annually, representing 25% of the reductions necessary to meet the CO2 reduction goal. In addition, nuclear energy could help stabilize the cost of electricity by reducing the United States' dependence on volatile natural gas prices. Although nuclear energy is currently more expensive than other forms of energy generation, the introduction of a carbon tax of just $5-$15 per megaton carbon would cause nuclear energy to become more competitive than coal and natural gas.

Investment in the expansion of nuclear energy was discontinued in the early 1990's because of concerns with safety, long-term storage of nuclear waste, high capital costs, and nuclear weapons proliferation. While these issues still drive current research, technological innovations are providing solutions and mitigation strategies to address these problems. This paper presents a probable course of development for the expanded use of nuclear power, considering the risks, benefits, costs, and technological feasibility of nuclear power generation.

New construction should begin immediately to supplement and replace the aging U.S. nuclear fleet. In the short term, improved Light Water Reactors employing an open, once-through or partially closed fuel cycle are commercially viable based on cost and technological readiness. However, issues with waste disposal and uranium availability make this option unsustainable in the long term. A fully closed fuel cycle can greatly improve energy utilization, increase long-term storage capacity, and help mitigate waste disposal hazards.

Since current fuel recycle technologies, such as PUREX, present unacceptable environmental safety and proliferation risks, new recycle processes that address these problems will need to be deployed. The time scales involved with developing some of these technologies (e.g. fast reactors) emphasize the need for immediate and extensive R&D efforts to secure the growth of nuclear energy resources for the future. Expansion of the sustainable, safe, clean and cost-effective electrical energy source presented by nuclear power is necessary to provide a foundation for future U.S. growth.

Notes:

1. "Carbon-free" in this context defines no carbon emissions due directly to energy generation and does not account for Life-Cycle related emissions in plant construction and materials.

2. One megaton equals one million metric tons.

See more of this Session: Chemical Engineering Advances in the Nuclear Fuel Cycle (including NED student competition)
See more of this Group/Topical: Nuclear Engineering Division

# Death of Energy

**Muhammad Shafiq Siraj**

Institute for Process and Particle Engineering

Graz University of Technology,

8010 Graz, Austria.

## Abstract

The main focus of this paper is on energy and its availability on the Earth, extending the energy boundaries beyond the Earth and looking deep into the future even after the life of the Sun. Implications of the second law of thermodynamics is also discussed in this paper. Secondly, the Sun-Dependent Energy Sources (SDES) and the Sun-Independent Energy Sources (SIES) especially geothermal and thermonuclear are analyzed in detail. An estimate of total reserves of these energy sources on the planet, and an optimal use is discussed in the last section of this paper enabling us to understand the energy issues on a real global scale. In the end, recommendations for a comprehensive global energy policy regarding the sequence of the use of SDES and SIES are given.

## Introduction

Energy is older than human or any other living specie on the Earth. It will not be wrong to quote that it is the energy which gave birth to all living species including human and is also the mother of all the matter in the Universe. Upon understanding the power and significance of the energy though in a limited perspective, ancient human started worshiping different forms of energy like fire, and energy sources like the Sun, which are still being worshiped in some parts of the world.

Sun is the source of all major forms of energy on the Earth e.g. solar, wind, hydro, and also the hydrocarbon sources of energy like coal, oil, and natural gas. It also contributes to the tidal energy from the oceans, in this case by its gravitational pull. The other two types of energies from the oceans which also depend on the Sun, and can be extracted, are: wave energy, and ocean thermal energy conversion. In ocean thermal energy conversion, the energy can be extracted by exploiting the temperature difference between the ocean waters on the surface and the ocean waters deep underneath.

However, there are some important energy sources on the Earth which are not or not primarily based/dependent on the Sun or solar energy. These are nuclear, thermonuclear (fusion), and geothermal, and can be treated as the Sun-Independent Energy Sources (SIES). The first one i.e. nuclear is the most controversial of all the energy sources being used, controlled energy from fusion is still not possible, and the third i.e. geothermal is limited in industrial extraction of energy for practical purposes and not feasible on most places of the planet with the existing technologies.

## Theory

Energy has no boundaries and the Earth receives as well as radiates back the energy from and to the Universe whereas the energy sources, their origin, strength, utilization, depletion rate, and exhaustion period have never been envisioned or evaluated on real global scale. They have been visualized mostly assuming the Sun as an infinite source of energy especially regarding its exhaustion time, thus giving rise to the misnomers like 'renewable energy sources'.

All nuclear fuels have certain half-lives, and no matter how long is the half-life, they will eventually exhaust. Geothermal energy source will also die out and so as the Sun which is the prime source of energy on the planet. We have never envisioned the exhaustion times for nuclear, geothermal, and all the Sun-Dependent Energy Sources (SDES) and a global energy policy keeping in view the energy is being constantly lost into the space with a constant increase in entropy or unavailable energy. Fortunately, we have Sun-independent energy sources in the form of geothermal, and light and stable nuclei suitable for fusion.

## Procedures

The procedure for calculations of these energy sources is based on law of conservation of energy, and the equation $E=mc^2$

## Results and Discussion

According to UN population data, the world population will stabilize just above 10 billion after the year 2200 [1] as shown in the fig.1.

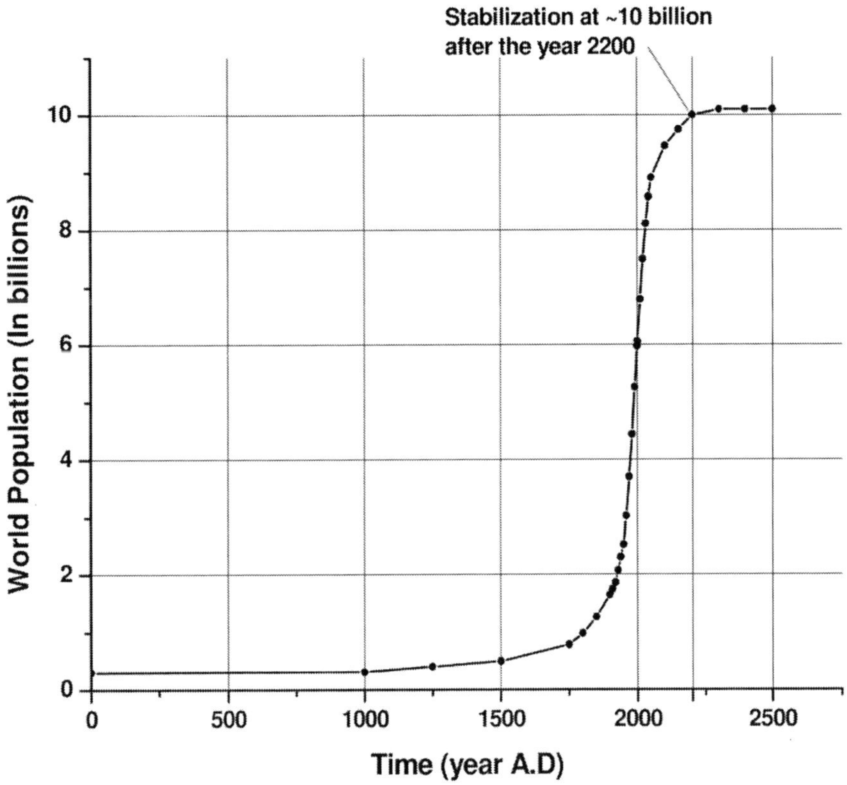


Fig.1. World Population

This baseline of 10 billion population is used for energy calculations from the year 2200 onwards. The reason for choosing year 2200 is because there are already a number of energy scenarios till the year 2100 and/or 2200, and to the best knowledge of the author there has never been a single effort for an energy scenario beyond the year 2200 on a real global scale.

Energy consumption per person in USA for the year 2008 [2] was ~10.94 kW. For the energy calculations a value of 10 kW per person will be used. However, the world average consumption per person is much lower than this value but it can be safely assumed that from the year 2200 onward, the use of energy by the rest of the world would increase but the overall consumption might not increase than the current average consumption of USA because of increase in the efficiency of all the energy producing and consuming sectors.

For a world population of 10 billion with an average power consumption of 10 kW per person, the total energy consumption by the world population from the year 2200 onwards comes out to be a near-constant value of 3154 EJ/year (100 TW). By this point of time, the two most controversial sources of energy i.e. fossils and nuclear would already have consumed up. The uranium resources estimated worldwide can fuel the nuclear reactors for at least next 100 years [3]. According to these new estimates, the conventional uranium resources are estimated to be around 5.5 million tonnes which can be mined economically (less than US$ 130/kg). Undiscovered resources which can be found based on current knowledge of geological characteristics of existing deposits are ~10.5 million tonnes [3]. Even if no further increase in demand of uranium is considered i.e. current uranium requirement stays constant, the nuclear power production would be wiped out before the year 2200.

Similarly, the fossil fuels including the current reserves, potential reserve growth, and undiscovered potential reserves of oil, gas, and coal would be depleted with an approximate exhaustion year of 2100 for oil, 2165 for gas and 2230 for coal [4].

Therefore, from the year 2200 we will be left with solar, hydro, wind, biomass, geothermal, and fusion.

Now, the potential share of each energy source is discussed.

Table 1a: Theoretical, Technical, and Economical Potential of Major SDES (Renewables) [a]

| Energy Resource | Theoretical Potential (EJ/year) | Technical Potential (EJ/year) | Availability (Billion Years from Now) |
|---|---|---|---|
| Hydro | 147 | 50 | 5 |
| Solar | $3.9 \times 10^6$ | >1575 | 8-10 |
| Wind | 6000 | 640 | 5-7 |
| Biomass | 2900 | >276 | 5 |
| Ocean [b] | 7400 | 74 | 5 |
| Total | $3.92 \times 10^6$ | >2615 | 5-10 |

[a] Source: WEA-2000, World Energy Assessment and Author's Calculations

Table 1b: Theoretical, Technical, and Economical Potential of Major SIES (Non-Renewables) excluding Fossils and Nuclear[a]

| Energy Resource | Theoretical Potential (EJ) | Technical Potential (EJ) | Availability (Years) |
|---|---|---|---|
| Geothermal | $140 \times 10^6$ (from 3-10 km only) | $28 \times 10^6$ | 17,755 (with 50% share of total requirement of 3154 EJ/year) |
| Fusion | $15.64 \times 10^{12}$ | $15.64 \times 10^{10}$ (with a conservative 1% of theoretical potential) | $5 \times 10^6$ (with 100 % share of total requirement of 31,540 EJ/year) |

[a] Source: The Future of Geothermal Energy, MIT Report 2006, and Author's Calculations

USA is estimated to have a geothermal energy of $14 \times 10^6$ EJ from a depth of 3 to 10 km [5]. 40% recovery is upper limit but even if a midrange of 20% recovery is considered, the total energy amounts to $2.8 \times 10^6$ EJ. A rough estimate for the world, as USA has ~5-10% of the world's geothermal energy, would give a value of $28 \times 10^6$ EJ recoverable geothermal energy for the world and a total potential of $140 \times 10^6$ EJ for a depth of only 3-10 km.

Among the four fusion reactions, D-T (Deuterium-Tritium) reaction is the easiest one followed by the D-D (Deuterium-Deuterium) reaction. Deuterium is in huge quantities on earth ($4.6 \times 10^{16}$ kg of Deuterium [6]).

New research [7] shows that 7.6 billion years from now, the Sun will turn to a red-giant, a first step towards its death before turning to a white and then black dwarf. The chance now is more that it will engulf the Earth when it will increase to 250 times of its present size as a red-giant. One possible escape is to use an asteroid to tug and pull the Earth from getting close enough to it. The chances are good that the mankind would have enough expertise and technology by that time to use an asteroid to pull the Earth away from the red-giant. The earth and the mankind can survive this gloomy fate only if they have enough energy sources at that time.

They should have enough energy sources for other various uses as well like;

- To use energy for making structures underground with cooling to avoid extreme heat at that time,
- shifting the necessary infrastructure to underground facilities including the fusion materials,
- energy for food growth underground,
- and for routine activities while staying underground until the Sun passes through the phase from red-giant to white dwarf.
- Lastly, bringing/developing the entire infrastructure back on the surface of the Earth once the surface is cooled down after experiencing a hot temperature.

The hot surface during this transition time can again be used for extraction of heat.

The real use of fusion materials for energy use will start after the Sun turns into a black dwarf. The energy will be required for starting a new life on the planet surface. Food growth can be made possible with the huge fusion reactors emitting thermal radiation to space and then reflected back to the surface of the Earth to provide artificial sunlight. The energy consumption per person may increase by a factor of 10 after this point i.e. 1000 kW per person or a total consumption of 31540 EJ/year.

## Conclusion

Within couple of centuries, we will be left with only SDES or the so-called renewables and SIES without nuclear and fossils. In the long term future policy of energy use, it is strongly recommended that share of SDES (including solar, hydro, and wind) be kept as maximum as possible because of three reasons; they are the cleanest energy sources, if not used they are lost forever (in the form of thermal radiation loss to space for solar and the lost kinetic energy for hydro and wind, leading to a constant increase of entropy or unusable energy), and thirdly because the fusion materials are the best long term energy storage solution for the future generations after the death of the Sun and the Sun-Dependent Energy Sources (SDES). At that time the energy would be required not only for routine activities but also for the food chain supply which otherwise depend on solar energy. Therefore, in the far future, the energy consumption per person would be much greater than the current estimated value of 10 kW for the year 2200 onward. The second energy source which has great potential and can be used along with SDES is geothermal and may be used between SDES and fusion because it offers an interim energy storage solution. Therefore, the sequence should be:

- first sources to utilize must be SDES with maximum share,
- second source should be geothermal with minimum to medium share, and
- third source should be fusion with a minimum share until the first source dies out.

## Nomenclature

EJ = exa joule ($1\times10^{18}$ joules)

kW = kilo watt ($1\times10^{3}$ watt)

TW =tera watt ($1\times10^{12}$ watt)

## References


[1] The World at Six Billion, ESA/P/WP.154, UN Population Division, October 1999.

[2] Annual Energy Review 2008, DOE/EIA-0384(2008), June 2009.

[3] Uranium 2007: Resources, Production and Demand, Joint Report by OECD NEA and IAEA, Paris, 2008.

[4] World Fossil Fuel Reserves and Projected Depletion, The Colorado River Commission of Nevada, March 2002.

[5] The Future of Geothermal Energy, MIT Report, 2006.

[6] Hamacher, T., and Bradshaw, A.M., "Fusion as a Future Power Source: Recent Achievements and Prospects", 18th World Energy Congress, October 2001.

[7] http://www.newscientist.com/article/dn13369-hope-dims-that-earth-will-survive-suns-death.html

# Computational Model of the Mark-IV Electrorefiner- 2D Potential and Current Distributions

Robert O. Hoover, Nuclear Engineering Program, University of Idaho, Idaho Falls, ID

Supathorn Phongikaroon, Department of Chemical Engineering and Nuclear Engineering Program, University of Idaho, Idaho Falls, ID

Michael F. Simpson, Pyroprocessing Technology Department, Idaho National Laboratory, Idaho Falls, ID

Tae-Sic Yoo, Pyroprocessing Technology Department, Idaho National Laboratory, Idaho Falls, ID

Shelly X. Li, Pyroprocessing Technology Department, Idaho National Laboratory, Idaho Falls, ID

An electrochemical process has been developed by Argonne National Laboratory (ANL) to treat sodium-bonded spent fuel from the Experimental Breeder Reactor-II (EBR-II) located at Idaho National Laboratory (INL). At the heart of this innovative process lies the Mark-IV (Mk-IV) electrorefiner (ER). The spent driver fuel from EBR-II is chopped into 1/4 inch segments and loaded into steel fuel dissolution baskets which act as the anode. Uranium and other active metals in the spent fuel are electrochemically dissolved while essentially pure uranium is transferred through a LiCl/KCl eutectic electrolyte and simultaneously deposited on a stainless steel cathode mandrel.

In order to better understand the processes occurring within the Mk-IV ER, a computational model has been developed based on fundamental electrochemical, thermodynamic, and kinetic theories using the *Matlab* computer software. Both the potential and current distributions throughout the ER can convey important information in better understanding the entire electrochemical process.

The potential distribution can be found using the Laplace equation: $\partial^2\Phi/\partial x^2 + \partial^2\Phi/\partial y^2 = 0$, where $\Phi$ is potential. The model determines the potentials at the anode and cathode throughout the electrorefining process. These values constitute Dirichlet boundary conditions at these locations, while at the ER outer wall, $\partial\Phi/\partial x + \partial\Phi/\partial y = 0$, or a Neumann boundary condition is used. Matlab has a partial differential equation (PDE) solver, which uses the finite element method, that has been combined with the model and used to solve this two-dimensional PDE.

The current distribution is then calculated from the resulting potential distribution. Assuming the bulk electrolyte is well mixed and there are no concentration gradients the current distribution can be described by: $i=-F^2*(\partial\Phi/\partial x + \partial\Phi/\partial y)*\Sigma(z_i^2*u_i*c_i)$, where i is current density ($A/m^2$), F is Faraday's constant (96,485 C/eq), $z_i$ is the ionic charge of species i (eq/mol), $u_i$ is species i's ionic mobility through the electrolyte ($(m^2\cdot mol)/(J\cdot s)$), and $c_i$ is the concentration of species i ($mol/m^3$).

Preliminary results show steep potential gradients ranging from 1-3 V/m in the regions directly between the anodes and cathode, while areas further from the electrodes have much less variable potential. The results reveal that the majority of the current, and therefore most of the uranium, travels through the area of highest potential gradient, or directly from the anodes to the cathode. Based on these preliminary results, the conductivity of the solution at average concentrations of ions, throughout the electrochemical process is approximately 193 mho/m. The two-dimensional potential and current distributions within the Mark-IV ER will be presented and analyzed for three different species: $U^{3+}$, $Pu^{3+}$, and $Zr^{4+}$ along with the total distributions. The results will also be shown for different times as the electrorefining process progresses.

See more of this Session: Chemical Engineering Advances in the Nuclear Fuel Cycle (including NED student

competition)
See more of this Group/Topical: Nuclear Engineering Division

competition)
See more of this Group/Topical: Nuclear Engineering Division

# Radiolabeling of Biomolecules with Cu-64 in Microfluidic Reactors

Amit V. Desai, Department of Chemical and Biomolecular Engineering, University of Illinois at Urbana-Champaign, Urbana, IL

Tobias D. Wheeler, Department of Chemical and Biomolecular Engineering, University of Illinois at Urbana-Champaign, Urbana, IL

Dexing Zeng, Radiological Sciences Division at Mallinckrodt Institute of Radiology, Washington University School of Medicine at St. Louis, St. Louis, MO

Mansi Thakkar, Department of Chemical and Biomolecular Engineering, University of Illinois at Urbana-Champaign, Urbana, IL

David E. Reichert, Radiological Sciences Division at Mallinckrodt Institute of Radiology, Washington University School of Medicine at St. Louis, St. Louis, MO

Paul J.A. Kenis, Department of Chemical and Biomolecular Engineering, University of Illinois at Urbana Champaign, Urbana, IL

Radiolabeled biomolecules are used for Positron Emission Tomography (PET), a medical diagnostic technique used to image tumors in the body, and as radiotherapeutic agents. The radiolabeled biomolecules used in these applications consist of a peptide or antibody coupled to a bi-functional chelator (BFC) that chelates a radioactive metal ion. Once introduced into the body, the biomolecule binds to receptors that are expressed on the surface of cancerous cells, thereby targeting the delivery of the radiometal. Among radiometals, copper 64 (Cu-64) is one of the most widely used isotopes for PET imaging, due to its simple and well-understood coordination and redox chemistry, and ability to form kinetically inert copper complexes for long term targeting and trapping (e.g. radiolabeled antibodies).

In conventional radiolabeling procedures, a large excess of the BFC (approximately 100 fold) is typically used for convenient handling of the chemicals and efficient radiolabeling. However, this excess results in lower specific activity of the radiolabeled probe, slower reaction times, and high consumption of expensive precursors. In order to improve the specific activity and image low abundance targets in vivo, extensive and time consuming separation procedures are required.

In this presentation, we will discuss the development of microfluidic reactors for efficient radiolabeling of biomolecules with Cu-64, and demonstrate higher specific activities than obtained in conventional, macro-scale processes. We will also demonstrate that the microfluidic approach allows for reduced consumption of expensive precursors, and synthesis of clinically relevant volumes (~50 ìL) and concentrations of solutions of radiolabeled biomolecules. The integration of various radiolabeling steps on a single system leads to improved automation of the radiolabeling process. We will conclude with a discussion on further potential improvements to the production of radiolabeled biomolecules through the application of microfluidic technologies.

See more of this Session: Chemical Engineering Advances in the Nuclear Fuel Cycle (including NED student competition)
See more of this Group/Topical: Nuclear Engineering Division

# Initial Production of Full-Scale Ceramic Waste Forms for the EBR-II Spent Fuel Treatment Process

**Michael F. Simpson**, Pyroprocessing Technology Department, Idaho National Laboratory, Idaho Falls, ID

Kenneth J. Bateman, Pyroprocessing Technology, Idaho National Laboratory, Idaho Falls, ID

Matthew Morrison, Pyroprocessing Technology, Idaho National Laboratory, Idaho Falls, ID

The Experimental Breeder Reactor-II (EBR-II) was a fast neutron spectrum test reactor that operated at Argonne National Laboratory-West (now Idaho National Laboratory) from 1963 until 1994. It employed sodium-bonded, metallic U-Zr fuel with a total heavy metal inventory at shutdown of approximately 25 metric tons. From 1996 to 1999, ANL successfully demonstrated at engineering-scale a pyrochemical process for treating the spent fuel that stabilized the sodium, recovered pure U metal, and generated durable high-level waste forms for the fission products, plutonium, and minor actinides. Since 1999, research and development has continued on the process to improve separation efficiency, lower operating cost, and develop advanced technologies to increase the functionality of the spent fuel treatment process. The process for disposing of waste salt from the electrorefiners is referred to as the ceramic waste process and has been scaled-up to achieve production operation capacity demands. The process includes zeolite size reduction, zeolite drying, salt/zeolite blending, and high temperature consolidation. Full-scale equipment has been installed in non-radiological laboratories at Idaho National Laboratory to initially make non-radioactive, full-scale waste forms to confirm the suitability of equipment design and to select operating parameters. In full-scale batches, zeolite-4A has been successfully dried to less than 0.5 wt% water. Also at this scale, salt has been very efficiently absorbed into the dried zeolite. Using the resulting salt-loaded zeolite, ceramic waste forms in excess of 300 kg have been made. From the initial lab-scale experiments, the waste form and the process by which it has been made has been scaled-up by a factor of 1500. The next step for process development is qualification of the process via repeatability runs to establish that given operating parameters will reliably produce identical results time after time. Once the process qualification is complete, final equipment installations will be completed in the hot cells, and production operations for the ceramic waste process with waste salt from the electrorefiners will commence.

See more of this Session: Chemical Engineering Advances in the Nuclear Fuel Cycle (including NED student competition)
See more of this Group/Topical: Nuclear Engineering Division

# Electrochemical and Photochemical Wastewater Treatment

Christoph Hoislbauer, University of Technology Graz, Department of Chemical Engineering and Environmental Technology, Graz, Austria

Matthäus Siebenhofer, Department of Chemical Engineering and Environmental Technology, Graz University of Technology, Graz, Austria

Photochemical and electrochemical Advanced Oxidation Processes (AOP/EAOP) are efficient oxidation operations for degradation of persistent pollutants such as pharmaceuticals, pesticides, substituted aromatic hydrocarbons, dyes and complexing agents. The UV/H2O2 -, UV/O3- processes and the Photo-Fenton process are representative photochemical AOPs. Anodic oxidation is an electrochemical AOP. Both technologies generate reactive hydroxyl radicals which mineralize contaminants or at least convert pollutants into less toxic or biodegradable products.

The aim of the project is the comparison and documentation of approved photochemical and electrochemical processes for different substance classes. This catalogue should help to evaluate a proper technology for specific application. Criterion for comparison is the potential of degradation and mineralization, as well as the specific energy consumption.

Several substances, the selection was mainly based on the properties of substituents, were investigated. The experiments were carried out with a start concentration of 500 ppm of the investigated substances. Photochemical degradation was carried out in a Heareus Noblelight UV-reactor system (UV-RS-2), equipped with a 15 W mercury low-pressure lamp. Electrochemical degradation was performed in open cell electrolysis with diamond coated electrodes and 0.05 M Na2SO4 for conductivity reasons. The results demonstrate different rate of mineralization and degradation related to the substituent of the model substance and the absorption properties in the UV Range. In general electrochemical experiments show a simultaneous degradation and mineralization whereas photochemical experiments show degradation of substances in a first step followed by mineralization.

See more of this Session: Advanced Oxidation/Reduction Applications: Liquid/Gas Phase I
See more of this Group/Topical: Environmental Division

# Mechanistic Study On the Electrochemical Decolorisation of Rhodamine 6G Using RuO2 Coated Ti DSA

Yu-Ming Zheng, Division of Environmental Science and Engineering, Natioanl University of Singapore, Singapore, Singapore

Rita Farida Yunus, Environmental Science and Engineering, National University of Singapore, Singapore, Singapore

**J. Paul Chen**, Division of Environmental Science and Engineering, National University of Singapore, Singapore, Singapore

In this study, the performance of electrochemical treatment of Rhodamine 6G containing water by using $RuO_2$ coated Ti mesh DSA is evaluated, as well as the mechanism of decolorisation process is explored. The effects of solution pH, temperature and addition of auxiliaries on the electrochemical process were examined. Results show solution pH and temperature play import roles during the electrochemical treatment of Rhodamine 6G containing water. Lower solution pH and lower temperature facilitate the decolorisation of the dye-containing water. The addition of auxiliaries dose not significantly effect the electrochemical decolorisation of the dye-containing water. Under optimal experimental condition, complete decolorisation of the dye-containing water can be obtained within 5 min. The electrochemical process can achieve 40.6 and 37.3% mineralization of Rhodamine 6G solution without and with auxiliaries within 5 min, respectively. It was found that water-insoluble foam product formation and mineralization of the Rhodamine 6G were involved during the electrochemical decolorisation of the dye-containing water. Total organic carbon, total nitrogen mass balance analyses, UV-Vis, FTIR and XPS spectroscopy were used to explore the mechanism of the electrochemical decolorisation process. UV-Vis spectra analysis shows the foam product has different molecular structure from Rhodamine 6G. FT-IR, XPS, carbon and nitrogen analyses indicate: amine and methyl groups are first detached from the dye molecule, which changes the dye from polar into non-polar and results in formation of water-insoluble substance; then partial of the water-insoluble substance are floated by the hydrogen bubble generated on cathode, and the others in the solution are further mineralized via aromatic opening and oxidizing. A conceptual model for the electrochemical decolorisation is proposed to illustrate the mechanism.

See more of this Session: Advanced Oxidation/Reduction Applications: Liquid/Gas Phase I
See more of this Group/Topical: Environmental Division

# Eliminating Contaminants From Drinking Water or Fluids Using Pulsed Corona Discharges

Chinyere P. Mbachu, Chemical Engineering, Tennessee Technological University, Cookeville, TN

Pedro E. Arce, Chemical Engineering, Tennessee Technological University, Cookeville, TN

Dennis George, Center for the Management, Utilization and Protection of Water Resources, Tennessee Tech University, Cookeville, TN

The use of high oxidation methods to eliminate contaminants from drinking water or fluids is an attractive and efficient approach. Earlier work (Sharma et al., 1993) has shown the elimination of highly toxic contaminants such as phenol and more recently fumic traces (Smith, 2004) and proteins from water fluents (Ibañez, 2004). Pulsed corona high oxidation can be enhanced by adding either oxygen, in order to produce high active radicals, or iron to increase the effectiveness by the presence of the photo-Fenton reaction.

In this research proposal, we are interesting in determining the efficiency of the method in application such as sanitation of drinking water or other related fluids such as fruit juices. In particular, the technique has a low heat generation and leads to negligible or moderated temperature increases. This option is highly desirable as an alternative to high temperature methods such as pasteurization. Both organic and biological contaminants will be tested experimentally to determinate the ability of pulsed corona in eliminating them and leading to a safe drinking product.

See more of this Session: Advanced Oxidation/Reduction Applications: Liquid/Gas Phase I
See more of this Group/Topical: Environmental Division

# Investigating the Photocatalytic Performance of Rapid Settling Pnipam-Titania Microcomposites

**Cecil Coutinho**, Chemical & Biomedical Engineering, University of South Florida, Tampa, FL

Vinay K. Gupta, Chemical & Biomedical Engineering, University of South Florida, Tampa, FL

Novel composites were prepared that were composed of titania (TiO2) nanoparticles embedded within cross-linked microgels of poly(N-isopropylacrylamide). Interpenetrating linear chains of poly(acrylic acid) were used to functionalize the nanoparticles of TiO2 for dispersal within the porous framework of the stimuli responsive microgels. The microcomposite particles showed rapid sedimentation, which is useful for the gravity separation of these particles in applications such as environmental remediation via photocatalytic degradation. The settling behavior of microcomposites was studied optically using turbidity measurenets. The settling of the polymer-titania composites occurred over minutes and was much faster than solid, impermeable spheres. As the content of TiO2 increased within the particles from 10wt% to 75wt%, the increased effective particle density led to significant decrease in the settling time. A simple mathematical model was used to interpret the optical measurements in terms of the distribution of settling velocities of the particles. These microcomposites were also very photocatalytically active and the degradation kinetics in aqueous solutions was investigated using methyl orange as a model contaminant. Kinetics of the photodegradation was evaluated by monitoring the decline in the methyl orange concentration using UV-Vis spectroscopy. Degradation of methyl orange using freely suspended titania (DegussaTM P25) was also conducted for comparison with the composites. The results reported here are a necessary input for the development of processes that could utilize composites for photocatalysis.

See more of this Session: Advanced Oxidation/Reduction Applications: Liquid/Gas Phase I
See more of this Group/Topical: Environmental Division

# Kinetic Modeling of Phenol Oxidation by Fenton's Reagent in Culture Media

**Pardeep Kumar**, Chemical Engineering, University of Saskatchewan, Saskatoon, SK, Canada

Hossein Nikakhtari, Chemical Engineering, University of Saskatchewan, Saskatoon, SK, Canada

Mehdi Nemati, Chemical Engineering, University of Saskatchewan, Saskatoon, SK, Canada

Gordon Hill, Chemical Engineering, University of Saskatchewan, Saskatoon, SK, Canada

Phenols are commonly found in the wastewater effluents of a variety of chemical industries. Higher concentrations of phenols are toxic and resistant to biological degradation. Advanced oxidation processes (AOPs) can be combined with biological processes to effectively remove higher concentrations of phenols. In this study, chemical oxidation of phenol has been investigated by Fenton's reagent in both pure water and culture media required for the purpose of microbial growth and bioremediation. The oxidation of phenol by Fenton's reagent shows that extent of conversion and the oxidation rate of phenol in the culture media are quite different from that reported for the same reaction in pure water.

Intermediate compounds, catechol, hydroquinone and maleic acid, have been experimentally identified and quantified during phenol oxidation by Fenton's reagent. A kinetic model for phenol degradation that includes the dynamic formation and destruction of these intermediate compounds has been developed. The final oxidized product is carbon dioxide. Experiments were conducted at different initial concentrations of Fenton's reagents in both pure water and culture media. The phenol removal increased from 62% to 100% with the increase in hydrogen peroxide dosage from 1000 mg/L to 5000 mg/L at an initial phenol concentration of 2000 mg/L. The kinetic model developed fits the experimental data reasonably well and is capable of predicting the dynamic concentrations of each intermediate compound. The comparison of kinetic constants in pure water and culture media will be shown.

See more of this Session: Advanced Oxidation/Reduction Applications: Liquid/Gas Phase I
See more of this Group/Topical: Environmental Division

# Sonochemical Removal of Nitric Oxide Using a Novel Ultrasonic CSTR

**Ibraheem R. Muhammad**, Chemical & Bioengineering Department, North Carolina Agricultural and Technical State University, Greensboro, NC

Nymul E. Khan, Chemical & Bioengineering Department, North Carolina Agricultural and Technical State University, Greensboro, NC

Naresh N. Mahamuni, Chemical & Bioengineering Department, North Carolina Agricultural and Technical State University, Greensboro, NC

Yusuf G. (Debo) Adewuyi, Chemical & Bioengineering Department, North Carolina Agricultural and Technical State University, Greensboro, NC

The combustion of fossil fuel releases a large amount of NOx and sulfur dioxide (SO2) into the atmosphere which is a major threat to human health and environment. NOx, together with SO2 is the major contributer to acid rain that harms forest crops and buildings, as well as aquatic life. NOx also constitutes one of the main ingredients involved in the formation of ground-level ozone and urban smog through photochemical reactions, which can trigger serious respiratory problems. The efficient, cost effective and environmentally friendlier removal of NOx and SO2 from flue gases poses a considerable industrial problem. Available control technologies such as Selective Catalytic Reduction (SCR) used for high NOX removal has very high capital cost and suffers from various problems including disposal of spent toxic catalysts. Selective noncatalytic reduction (SNCR, urea injection) is highly temperature dependent and SNCR, ammonia injection (Thermal DeNOX) has moderately high capital cost and ammonia handling problems. Low-NOX burners (LNB) also has moderately high capital cost. Commercial applications of methods involving the chemical conversion of NO into soluble NO2 involving numerous liquid adsorbents in various gas-liquid contactors have been slow due to their complexities and additional costs of chemicals. Thus, there is need in the aqueous scrubbing arena for environmentally friendly and cost-effective alternatives for comprehensive treatment of NOx from industrial flue gases. Our research involves the aqueous removal of NOx (NO and NO2) from flue gases induced by ultrasound or hydrodynamic cavitation. Cavitation is the formation, growth and implosive collapse of gas- or vapor-filled microbubbles and can be induced acoustically or hydrodynamically in a body of liquid. Acoustic or ultrasonic cavitation involves the formation and subsequent collapse of microbubbles from acoustical wave-induced compression/rarefaction. On the microscale, the collapse of the bubbles leads to surprisingly high estimated local temperatures and pressures up to and above 5000 C and 500 atmospheres, respectively. These rather extreme conditions are very short lived, but have been shown to produce several highly reactive species in the aqueous systems including hydroxyl (•OH), hydrogen (H•) and hydroperoxyl (HO2•) radicals, and hydrogen peroxide.

In previous studies, we demonstrated that the absorption of NO (50-1040 ppm) into water with simultaneous oxidation to nitrite and nitrate induced by ultrasonic irradiation at a fixed frequency of 20 kHz could be accomplished using a sonochemical bubble column reactor in the absence and presence of SO2 (52-4930 ppm) and NaCl at room temperature (23 $\pm$ 2oC). The fractional conversions of NO were found to range from 60 to 85% while complete removal of SO2 was observed for all the inlet gas concentrations studied. The fractional conversions of NO were found to range from 60 to 85% while complete removal of SO2 was observed for all the inlet gas concentrations studied. It was shown that the presence of about 2520 ppm SO2 in combination with 0.01 M NaCl further enhanced NO removal. Combinative effect of sonication and chemical oxidation using 0.005-0.05M oxone was also studied. It was demonstrated that lower concentrations of HSO5- (0.005 M) enhanced NO removal efficiency. This paper will present new results obtained using our one-of it-kind sonochemical gas-liquid continuous stirred tank reactor,newly designed and built to provide drastic kinetic rate enhancement for demanding operations.

References

(1) Adewuyi, Y.G., Owusu, S.O. Ultrasound-induced Aqueous Removal of Nitric Oxide from Flue Gases. Effects of Sulfur Dioxide, Chloride and Chemical Oxidant. Journal of Physical Chemistry A. 2006. 110, 11098-11107.

(2) Owusu, S.O; Adewuyi, Y.G. Sonochemical Removal of Nitric Oxide from Flue Gases. Industrial & Engineering Chemistry Research. 2006, 45, 4475-4485.

See more of this Session: Advanced Oxidation/Reduction Applications: Liquid/Gas Phase II
See more of this Group/Topical: Environmental Division

# MODELING THE SONOCHEMICAL REMOVAL OF NITRIC OXIDES USING THE RAYLEIGH-PLESSET-NOLTINGK-NAPPIRAS-PORITSKY (RPNNP) BUBBLE DYNAMICS EQUATION

*Nymul E. Khan and Yusuf G. Adewuyi, Chemical and Bioengineering Department, North Carolina A&T State University, Greensboro, NC*

## Introduction

$NO_x$ (NO & $NO_2$) and $SO_2$ – mainly generated from combustion of fossil fuel in power plants and automobiles – have severe detrimental effects on plants, animals and humans. Numerous environmental and health effects such as acid rain, formation of ground-level ozone, eutrophication, respiratory and heart diseases in humans are caused by $NO_x$. Although emissions of $NO_2$ and $SO_2$ can be controlled easily by scrubbing, nitric oxide (NO) is not easily removed by aqueous scrubbing due to its very low solubility in such solutions. Ultrasound has been known to enhance the rate of many chemical reactions. Owusu and Adewuyi previously reported the effectiveness of sonochemical methods for the removal of NO.[1-2]
Not many studies in the field of ultrasound have focused on gas absorption. The few that have are mostly mass-transfer studies [3]. There has been no investigation studying the applicability of ultrasound in the absorption of a commercially important gas. In this perspective our previous studies have been solely about the removal nitric oxide (NO) from flue gases using a sonochemical bubble column reactor [1-2]. Absorption and conversion in the range of 65-80% have been shown to be possible.

In the current paper the development of a mathematical model for this system has been undertaken for the understanding of the different mechanisms involved. We propose a model to relate the cavitation bubble dynamics phenomena with the reaction in liquid phase and the simultaneous transfer of mass from the gas to the liquid phase to predict the rate of absorption in presence of ultrasound. We have developed a comprehensive model which takes into consideration (i) dynamics of cavitation, (ii) generation of oxidizing species from bubble collapse through reaction kinetics, (iii) transfer of NO from from gas to liquid, (iv) chemical reactions of oxidizing species with dissolved NO.
Sonochemical techniques utilize ultrasound to produce an oxidative environment via acoustic cavitation due to the formation and subsequent collapse of microbubbles from acoustical wave-induced compression/rarefaction. Ultrasound creates alternating compression and rarefaction pressure fields in a liquid. This causes tiny gas bubbles trapped in the liquid or in the crevices of surfaces to expand and then violently collapse. During the expansion phase water evaporates into the bubble and condenses during the collapse [4-7]. However, since the collapse is extremely fast and almost adiabatic in nature most of the vapor trapped inside the bubble cannot escape and is disintegrated due to the very high temperature reached inside the bubble (≈ 3000-5000 K) [8-10] to form reactive species such as hydroxyl (·OH), hydroperoxyl (·$HO_2$) radicals or oxygen (O) and hydrogen (H) atoms and hydrogen peroxide ($H_2O_2$) [11]. These species diffuse out of the bubble into a thin interfacial layer where they either recombine or react with compounds present in the liquid.

Majority of the studies for the application of sonochemistry are in the field of wastewater treatment [12-21]. The mechanisms related to the reaction-enhancement by cavitation have been elucidated in the many studies done on the sonochemical degradation of hazardous organic compounds found in wastewater [12-19]. This system is different from ours in terms of the number and types of phases involved but we can certainly expect that the mechanisms that lead to the chemical effects of cavitation to remain the same. Three different mechanisms have been proposed to explain the degradation of organic pollutants in wastewater. The first is the pyrolysis in the cavitation bubble interior. Volatile solutes evaporate into the bubble during the expansion phase and during the collapse the extremely high temperatures reached break them down into other species and thus causing their removal. The second mechanism is the hydroxylation of the species in the interfacial region by the ·OH radicals generated in the bubble. The H atom does not penetrate very far outside the bubble [20] and therefore the ·OH radical is the more prominent species and it is also one with the most oxidative potential of any known oxidative species except fluorine. The third possible mechanism, proposed by Hua et. al. [21], is the formation of super-critical water in a small shell around the bubble which promotes the reactions.

Earlier, studies were done on the degradation of gases in aqueous solutions [22-29] and important conclusions about the mechanism of sonochemistry were drawn from these studies. The primary mechanism for the decomposition of gases by ultrasound appears to be pyrolysis in the cavitation bubbles with hydroxylation in the interfacial sheath providing secondary pathways. But in their study using acetylene Hart et. al. [28] found out that the dominant mechanism shifts between that of pyrolysis in the bubble core and that of hydroxylation by ·OH radicals at the interface. They concluded that the lower concentration favors reaction pathways in the interfacial zone. Kotronarou and Hoffmann [29] proposed a model for the aqueous phase oxidation of $H_2S$ by ultrasound. A liquid phase reaction mechanism involving ·OH radicals generated by cavitational collapse was used in their model. By fitting with experimental data they have estimated the uniform release of ·OH radicals into the bulk solution to be of the order of 3.5 μM/min and a steady state concentration ≤ 0.1 μM. In view of these studies it is assumed that NO is destroyed mainly through reaction in the interfacial zone of the cavitation bubble with the OH radicals generated from the bubble interior. Hence, it would be assumed that the NO is removed primarily by attack of OH radicals in the interfacial zone of the cavitation bubble.

## Cavitation model

### *Bubble dynamics*

The motion of the bubble wall is described by the Keller equation [30], which is a modification of the well known Rayleigh-Plesset-Noltingk-Nappiras-Poritsky (RPNNP) equation and takes into account the liquid compressibility.

$$\left(1-\frac{\dot{R}}{c_\infty}\right)R\ddot{R}+\frac{3}{2}\dot{R}^2\left(1-\frac{\dot{R}}{3c_\infty}\right)=\frac{1}{\rho_L}\left(1+\frac{\dot{R}}{c_\infty}\right)\left(P_g-P_s(t)-P_\infty\right)+\frac{R}{c_\infty\rho_L}\frac{dP_g}{dt}-4\eta\frac{\dot{R}}{R}-\frac{2\sigma}{R} \quad (1)$$

where · denotes the time derivative (d/dt), $R$ is the radius of the bubble, $c_\infty$ is the speed of sound in the liquid, $P_g$ is the pressure of the gas in the bubble, $P_s$ is the time varying pressure field, and $P_\infty$ is the hydrostatic pressure, $\sigma$, $\mu$, $\rho_L$ denotes the surface tension, viscosity and density of the liquid respectively.

### *Differential mole balance of water vapor (transport and chemical reactions)*

The total rate of change of water vapor in the bubble is given by the following expression.

$$\frac{dn_w}{dt}=4\pi R^2\frac{10^3N_A}{M_{H2O}}\dot{m}+\frac{4}{3}\pi R^3\sum_j \alpha_{H_2O,j}r_j \quad (2)$$

where $n_w$ is the number of molecules of water in the bubble, $M_{H_2O}$ is the molar mass of water, $N_A$ is the Avogadro's constant. The second term represents the consumption by chemical reaction where $\alpha_{H_2O,j}$ is the stoichiometric coefficient of water in and $r_j$ is the rate expression of $j^{th}$ reaction.

$\dot{m}$ is the flux of water vapor across the bubble boundary and is given by:

$$\dot{m}=\dot{m}_{eva}-\dot{m}_{con} \quad (3)$$

$$\dot{m}_{eva}=\frac{\alpha}{\sqrt{2\pi R_g}}\left(\frac{p^*(T_L)}{\sqrt{T_L}}\right) \quad (4)$$

$$\dot{m}_{con}=\frac{\alpha}{\sqrt{2\pi R_g}}\left(\frac{\Gamma p_v}{\sqrt{T}}\right) \quad (5)$$

where $\alpha$ is the mass accommodation coefficient (literature suggests a value between 0.17 to 0.32 [31] so we used a conservative value of 0.17), $p^*$ is the saturated vapor pressure and $p_v$ is the partial pressure of water in the bubble. The correction factor, $\Gamma$, is given by eq. 6.

$$\Gamma=\exp(-\Omega^2)-\Omega\sqrt{\pi}\left(1-\frac{2}{\sqrt{\pi}}\int_0^\Omega \exp(-x^2)\,dx\right) \quad (6)$$

$$\Omega=\frac{\dot{m}}{p_v}\left(\frac{R_vT}{2}\right)^{\frac{1}{2}} \quad (7)$$

***Differential mole balance of other species (chemical reactions)***

The chemical rate of change of species $i$ is given by the sum of generation/consumption from all elementary reactions involving that species with their proper stoichiometric weight $\alpha_{i,j}$

$$\left(\frac{dn_i}{dt}\right)_c = V \times \sum_j \alpha_{i,j} r_j \quad (8)$$

A total of 49 reactions involving 19 species have been taken into account [32]. Each of these elementary reactions has a forward and backward rate per unit volume, $r_{f,j}$, $r_{b,j}$, which is described by means of modified Arrhenius laws.

$$r_{f,j} = A_{f,j} T^{B_{f,j}} \exp\left(-\frac{C_{f,j}}{kT}\right) \prod_i N_i^{\upsilon_{j,i}} \quad (9)$$

$$r_{b,j} = A_{b,j} T^{B_{b,j}} \exp\left(-\frac{C_{b,j}}{kT}\right) \prod_i N_i^{\upsilon_{j,i}} \quad (10)$$

The constants $A_j$, $B_j$, $C_j$ are constants corresponding to $jth$ reaction. The reactions are taken from ref. [32]. The overall rate is given by the difference of forward and backward rate

$$r_j = r_{f,j} - r_{b,j} \quad (11)$$

***Energy balance***

The temperature change inside the bubble can be obtained using the following equation.

$$\frac{dT}{dt} \sum_i \frac{\partial e_{th,i}}{\partial T} n_i = 4\pi R^2 \kappa \left.\frac{\partial T}{\partial r}\right|_{r=R} + -4\pi R^2 P_g \frac{dR}{dt} + \left(h_w - e_{th,w}\right) \frac{dn_w}{dt} + \sum_i \left(\frac{dn_i}{dt}\right)_c \Delta H_i^f \quad (12)$$

where $T$ represents the temperature inside the bubble, $e_{th,i}$ denotes the thermal energy and $e_{form,i}$ denotes the enthalpy of formation of species $i$, $n_i$ is the number of moles of species $i$, $\kappa$ is the thermal conductivity of the gas, $h_w$ is the molecular enthalpy of water vapor at the cold bubble wall, $e_{form,w}$ is the enthalpy of formation of water vapor.

The value of $e_{th,i}$ is given by the expression [33]:

$$e_{th,i} = \left(\frac{f_i}{2} kT + \sum_l \frac{k\Theta_{i,l}}{\exp\left(\frac{\Theta_{i,l}}{T}\right) - 1}\right) \quad (13)$$

and $f_i$, $\Theta_{i,l}$, $l = 1, \ldots\ldots, l_{max}$, denotes the number of translational and rotational degrees of freedom and the characteristic vibrational temperatures of species $i$, the values are obtained from ref. [33].

## Extrapolation of cavitation model to a sonochemical reactor

The above model for a single bubble furnishes information about the amount of OH radicals generated at the collapse and the change of bubble volume over one acoustic cycle. Using these two quantities we can determine on an order of magnitude the values of the range of the number of cavitation bubbles in a sonochemical reactor and subsequently the rate of generation of the OH radical. The mass transfer of NO from gas bubbles entering the reactor together with the following reaction scheme (Table 1) in the liquid phase is then used to predict the rate of removal of NO.

**Table 1.** Reaction scheme for the consumption of NO in the liquid phase

| No. | Reaction | k (L mol[-1] s[-1]) | Ref. |
|---|---|---|---|
| 1 | $NO + OH \rightarrow H^+ + NO_2^-$ | $2x10^{10}$ | [34] |
| 2 | $OH + OH \rightarrow H_2O_2$ | $5.5x10^9$ | [34] |
| 3 | $OH + H_2O_2 \rightarrow O_2^- + H_2O$ | $2.4x10^7$ | [34] |
| 4 | $O_2^- + OH \rightarrow OH^- + O_2$ | $7x10^9$ | [34] |
| 5 | $OH + OH^- \rightarrow O.^- + H_2O$ | $1.3x10^{10}$ | [34] |
| 6 | $NO_2^- + OH \rightarrow NO_2 + OH^-$ | $1x10^{10}$ | [34] |
| 7 | $NO_2 + OH \rightarrow HNO_3$ | $1x10^{10}$ | [34] |

| | | | |
|---|---|---|---|
| 8 | $O_2^{\cdot -} + NO \rightarrow ONOO^-$ | $4.3 \pm 0.5x10^9$ | [34] |
| 9 | $O.^- + H_2O \rightarrow OH^- + OH$ | $1.7x10^6$ | [34] |
| 10 | $O.^- + OH \rightarrow HO_2^-$ | $\leq 2x10^{10}$ | [34] |
| 11 | $H_2O_2 \leftrightarrow H^+ + HO_2^-$ | pK = 11.65 | [29] |
| 12 | $HNO_2 \leftrightarrow H^+ + NO_2^-$ | pK = 3.15 | [29] |
| 13 | $HO_2 \leftrightarrow H^+ + O_2^{\cdot -}$ | pK = 4.46 | [29] |

Yasui [35] has suggested an expression to calculate the rate of dissolution of the different species in the liquid using uptake coefficient ($\Theta$).

$$r_{d,i} = \Theta\sqrt{\frac{kT_B}{2\pi m_i}\frac{n_i}{V}} \times 4\pi R^2 \quad (14)$$

where $m_i$ is the molecular mass of the species.

The time-averaged rate of dissolution of OH radicals over one acoustic cycle from a single bubble is,

$$\langle r_{d,OH}(R_0)\rangle_t = \frac{\int_0^{t_{cycle}} r_{d,OH}(R_0,t)dt}{t_{cycle}} \quad (15)$$

where $\langle\ \rangle_t$ represents averaged quantity over $t$.

The average volume of a bubble over one acoustic cycle is,

$$\langle V(R_0)\rangle_t = \frac{\int_0^{t_{cycle}} V(R_0,t)dt}{t_{cycle}} \quad (16)$$

The number of bubbles has been estimated from the void fractions calculated in Tsochatzidis et. al. [36].

$$N = \frac{\epsilon}{\langle V(R_0)\rangle_t} \quad (17)$$

where $\epsilon$ is the void fraction and $N$ is the number of bubbles per unit volume.

Once this is calculated the average rate of generation of OH radicals in the reactor can be given by the expression,

$$S_{OH} = \langle r_{d,OH}(R_0)\rangle_t N \quad (18)$$

## Results and Discussion

### *Simulation of cavitation dynamics*

The bubble dynamics equations have been used to generate information such as bubble volume vs. time and no. of reactive radicals vs. time. These information allow us to calculate the number of bubbles and the rate of generation of OH radicals per unit time per unit volume. The illustrative results from the bubble dynamic simulations are shown in Figs. 1-4. The conditions used for these simulations are taken from Didenko and Suslick [37] in order to be able to validate our results against theirs. For the validation, we have compared our results against those found experimentally by Didenko and Suslick [37] and those from a more detailed model by Yasui et. al. [35].

The following are the bubble dynamics simulation results.

Conditions: f = 52 kHz, $P_A$ = 1.52 bar, $T_{amb}$ = 276 K, $P_h$ = 1 atm, $R_0$ = 4 μm.

**Table 2.** Comparison of our results with other models and experiment

| | Experiment [37] | Yasui et. al. [35] | Diffusion limited [35] | Present 4<br>95% Ar, 2.5% N2 and 2.5% O2 |
|---|---|---|---|---|
| $T_{max}$ | ... | 10900 K | 9400 K | 10200 K |
| Photon | $7.5x10^4$ | $8.0x10^4$ | $4.9x10^4$ | ... |
| | | | | |
| OH | $8.2x10^5$ | $6.6x10^5$ | $9.2x10^5$ | $1.1841x10^5$ |
| O | ... | $1.3x10^7$ | $1.1x10^7$ | $1.9562x10^7$ |

| | | | | |
|---|---|---|---|---|
| $H_2O_2$ | ... | 6.3x10$^6$ | 9.2x10$^6$ | 2.6197x10$^6$ |
| $O_3$ | ... | 3.4x10$^4$ | 4.0x10$^4$ | 3.1830x10$^4$ |
| $HNO_2$ | 9.9x10$^6$ | 2.3x10$^6$ | 2.0x10$^6$ | 6.3503x10$^1$ |
| $HNO_3$ | ... | 8.4x10$^5$ | ... | 2.107710$^4$ |

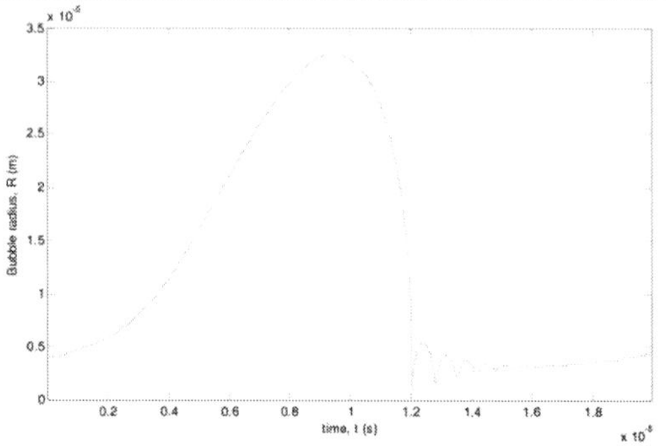


**Figure 1.** Cavitation bubble radius vs. time (1 acoustic cycle)

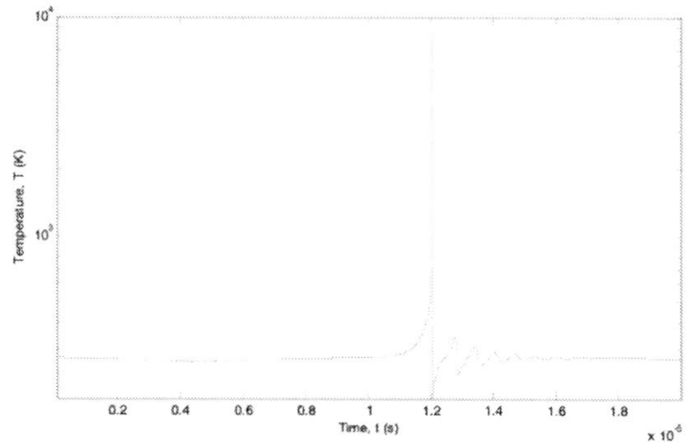


**Figure 2.** Temperature inside the cavitation bubble vs. Time (1 acoustic cycle)

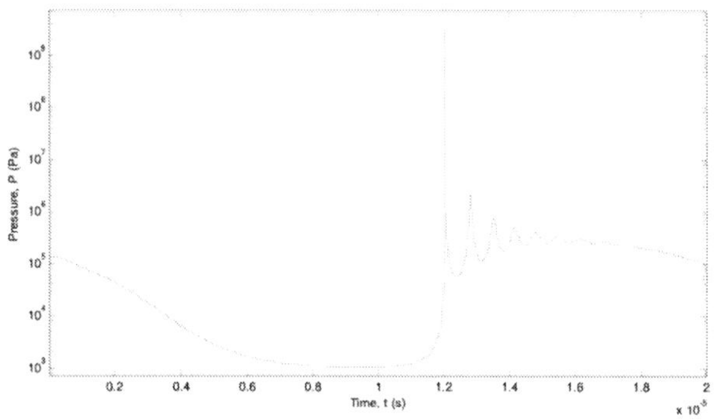


**Figure 3.** Pressure inside a cavitating bubble vs. Time (1 acoustic cycle)

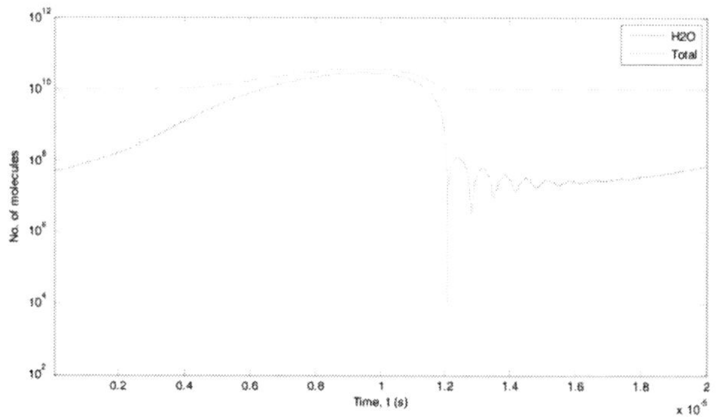


**Figure 4.** No. of molecules inside a cavitating bubble vs. Time (1 acoustic cycle)

This cavitation model applied to the conditions used in our experiments (f = 20 kHz, $P_A$ = 2 bar, $T_{amb}$ = 293 K, $P_h$ = 1 atm, $R_0$ = 5 μm) yields the following values:
The average volume of a cavitation bubble,

$$\langle V(R_0)\rangle_t = 3.51 \times 10^{-12}\ m^3$$

and rate of dissolution of OH radicals,

$$\langle r_{d,OH}(R_0)\rangle_t = 8.33 \times 10^{10}\ molecules/s$$

Therefore, the range of number of bubbles:

$$N = 2.85 \times 10^7\ to\ 1.42 \times 10^9\ bubbles/m^3$$

and the corresponding rate of generation of OH radicals:

$$S_{OH} = 8.14 \times 10^{-9}\ to\ 4.07 \times 10^{-7} \frac{mol}{L.s}$$

***Simulations for NO absorption***

Simulation conditions:
We are simulating a semibatch system for the absorption of NO. The gas phase is flowing continuously and the liquid phase is constant, with the ultrasonic probe dipped into it. The following are the global variables

that are used for the simulations corresponding to the experiments. They remain constant for all the simulations.

Gas flow rate, $Q^g = 0.1\ slpm = 1.667\frac{cm^3}{s}$

Liquid volume, $V^l = \frac{\pi}{4}D^2h = \frac{\pi}{4}x10^2x16\ = 1256.64\ cm^3$

Gas holdup, $V^g = \frac{\pi}{4}D^2h = \frac{\pi}{4}x5^2x1\ = 19.63\ cm^3$ (assuming a height increment of 1 cm)

Henry's law constant for NO, $H_{NO} = 5.1546\ x\ 10^5\ atm/(mol/cm^3)$

Physical mass transfer coefficient, $k_L a = 1.57x10^{-2}\ cm/s$ (calculated following ref. [38])

Following is an experimental data from NOx absorption studies (Fig. 5). The steady state conversion of NO is about 78%.

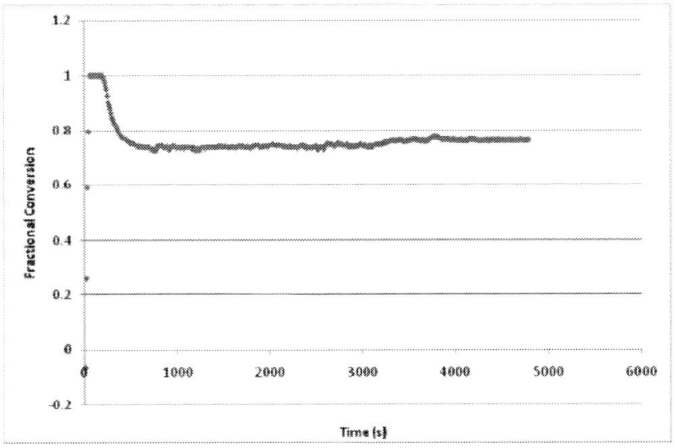


**Figure 5.** Experimental data for a NOx absorption study

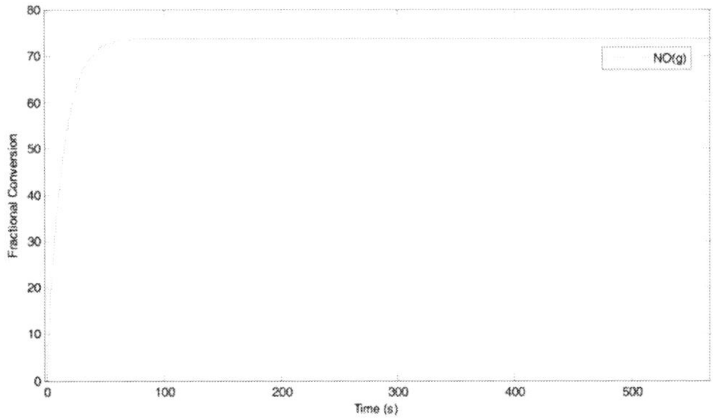


**Figure 6.** Simulation result of NOx absorption

In order to fit the experimental data, the simulation has been carried out at different mass transfer coefficient keeping the other parameters constant, since the mass transfer coefficient is the only parameter that is not known under ultrasonic irradiation. Following is the result of such a simulation (Fig. 6).

It was seen that around 75% conversion of NO is obtained at a mass transfer coefficient that is 5 times the physical mass-transfer coefficient calculated before. Hence, it is concluded that ultrasonic irradiation, apart from generating OH radicals which react with the dissolved NO to remove it, also increases the mass transfer at least 5 fold.

## Conclusion

A sonochemical bubble column for the absorption of NO was modeled. Bubble dynamics simulations of acoustic cavitation, incorporating liquid compressibility, mass and heat transport and chemical reaction, was performed. Coupling the chemistry generated from these microscopic phenomena to the reaction occurring in the liquid phase and mass transfer from the gas phase, the fractional conversion of NO was obtained. The results of the simulations indicate that only sonochemical oxidation cannot explain the previously reported results of NO removal by ultrasonic cavitation. Further simulations, with higher mass-transfer coefficients (equivalent to those found by other investigators in presence of ultrasound), show that only a combination of chemical activity and mass transfer enhancement by ultrasound can explain the results obtained.

## References


1. Adewuyi, Y. and S. Owusu, *Ultrasound-induced aqueous removal of nitric oxide from flue gases: Effects of sulfur dioxide, chloride, and chemical oxidant.* J. Phys. Chem. A, 2006. **110**(38): p. 11098-11107.
2. Owusu, S. and Y. Adewuyi, *Sonochemical removal of nitric oxide from flue gases.* Ind. Eng. Chem. Res, 2006. **45**(13): p. 4475-4485.
3. Laugier, F., et al., *Ultrasound in gas–liquid systems: Effects on solubility and mass transfer.* Ultrasonics-Sonochemistry, 2008.
4. Suslick, K., *Sonochemistry.* Science, 1990. **247**(4949): p. 1439-1445.
5. Adewuyi, Y., *Sonochemistry: environmental science and engineering applications.* Ind. Eng. Chem. Res, 2001. **40**(22): p. 4681-4715.
6. Mason, T. and J. Lorimer, *Applied sonochemistry.* 2002: Wiley-VCH Weinheim.
7. Thompson, L. and L. Doraiswamy, *Sonochemistry: science and engineering.* Ind. Eng. Chem. Res, 1999. **38**(4): p. 1215-1249.
8. Suslick, K., D. Hammerton, and R. Cline, *Sonochemical hot spot.* Journal of the American Chemical Society, 1986. **108**(18): p. 5641-5642.
9. Ciawi, E., et al., *Determination of temperatures within acoustically generated bubbles in aqueous solutions at different ultrasound frequencies.* J. Phys. Chem. B, 2006. **110**(27): p. 13656-13660.
10. Kanthale, P., M. Ashokkumar, and F. Grieser, *Estimation of Cavitation Bubble Temperatures in an Ionic Liquid.* 2007.
11. Makino, K., M. Mossoba, and P. Riesz, *Chemical effects of ultrasound on aqueous solutions. Formation of hydroxyl radicals and hydrogen atoms.* The Journal of physical chemistry, 1983. **87**(8): p. 1369-1377.
12. Hoffmann, M., I. Hua, and R. Höchemer, *Application of ultrasonic irradiation for the degradation of chemical contaminants in water.* Ultrasonics-Sonochemistry, 1996. **3**(3): p. 163-172.
13. Weavers, L., F. Ling, and M. Hoffmann, *Aromatic compound degradation in water using a combination of sonolysis and ozonolysis.* Environ. Sci. Technol, 1998. **32**(18): p. 2727-2733.
14. Hung, H., F. Ling, and M. Hoffmann, *Kinetics and mechanism of the enhanced reductive degradation of nitrobenzene by elemental iron in the presence of ultrasound.* Environ. Sci. Technol, 2000. **34**(9): p. 1758-1763.
15. Jiang, Y., C. Petrier, and T. Waite, *Effect of pH on the ultrasonic degradation of ionic aromatic compounds in aqueous solution.* Ultrasonics-Sonochemistry, 2002. **9**(3): p. 163-168.
16. Hua, I. and M. Hoffmann, *Kinetics and mechanism of the sonolytic degradation of CCl4: intermediates and byproducts.* Environ. Sci. Technol, 1996. **30**(3): p. 864-871.
17. Kotronarou, A., G. Mills, and M. Hoffmann, *Ultrasonic irradiation of p-nitrophenol in aqueous solution.* The Journal of physical chemistry, 1991. **95**(9): p. 3630-3638.
18. Tauber, A., H. Schuchmann, and C. von Sonntag, *Sonolysis of aqueous 4-nitrophenol at low and high pH.* Ultrasonics-Sonochemistry, 2000. **7**(1): p. 45-52.
19. Sivakumar, M., P. Tatake, and A. Pandit, *Kinetics of p-nitrophenol degradation: effect of reaction conditions and cavitational parameters for a multiple frequency system.* Chemical Engineering Journal, 2002. **85**(2-3): p. 327-338.
20. Hart, E. and A. Henglein, *Sonochemistry of aqueous solutions: H2-O2 combustion in cavitation bubbles.* Physical Chem, 1987. **91**(13): p. 3654-6.
21. Hua, I., R. Hoechemer, and M. Hoffmann, *Sonolytic hydrolysis of p-nitrophenyl acetate: the role of supercritical water.* The Journal of physical chemistry, 1995. **99**(8): p. 2335-2342.
22. Suslick, K., et al., *Alkane sonochemistry.* The Journal of physical chemistry, 1983. **87**(13): p. 2299-2301.
23. Fischer, C., E. Hart, and A. Henglein, *Ultrasonic irradiation of water in the presence of oxygen 18, 18O2: isotope exchange and isotopic distribution of hydrogen peroxide.* The Journal of physical chemistry, 1986. **90**(9): p. 1954-1956.

24. Fischer, C., E. Hart, and A. Henglein, *Hydrogen/deuterium isotope exchange in the molecular deuterium-water system under the influence of ultrasound.* The Journal of physical chemistry, 1986. **90**(2): p. 222-224.
25. Hart, E. and A. Henglein, *Sonolysis of ozone in aqueous solution.* The Journal of physical chemistry, 1986. **90**(14): p. 3061-3062.
26. Hart, E. and A. Henglein, *Sonolytic decomposition of nitrous oxide in aqueous solution.* The Journal of physical chemistry, 1986. **90**(22): p. 5992-5995.
27. Hart, E., C. Fischer, and A. Henglein, *Isotopic exchange in the sonolysis of aqueous solutions containing deuterium and methane.* Journal of Physical Chemistry, 1987. **91**(15): p. 4166-4169.
28. Hart, E., C. Fischer, and A. Henglein, *Pyrolysis of acetylene in sonolytic cavitation bubbles in aqueous solution.* Journal of Physical Chemistry, 1990. **94**(1): p. 284-290.
29. Kotronarou, A. and M.R. Hoffmann, *The chemical effects of collapsing cavitation bubbles: Mathematical modeling*, in *Advances in Chemistry series 244*, C.P. Huang, C.R. O'Melia, and J.J. Morgan, Editors. 1995, American Chemical Society: Washington DC.
30. Prosperetti, A. and A. Lezzi, *Bubble dynamics in a compressible liquid. Part 1. First-order theory.* Journal of Fluid Mechanics Digital Archive, 2006. **168**: p. 457-478.
31. Li, Y., et al., *Mass and thermal accommodation coefficients of H2O (g) on liquid water as a function of temperature.* J. Phys. Chem. A, 2001. **105**(47): p. 10627-10634.
32. Yasui, K., et al., *Theoretical study of the ambient-pressure dependence of sonochemical reactions.* The Journal of Chemical Physics, 2003. **119**: p. 346.
33. Toegel, R. and D. Lohse, *Phase diagrams for sonoluminescing bubbles: A comparison between experiment and theory.* The Journal of Chemical Physics, 2003. **118**: p. 1863.
34. Buxton, G., et al., *Critical review of rate constants for reactions of hydrated electrons, hydrogen atoms and hydroxyl radicals(•OH/•O ) in aqueous solution.* Journal of Physical and Chemical Reference Data, 1988. **17**(2): p. 513-886.
35. Yasui, K., et al., *Theoretical study of single-bubble sonochemistry.* The Journal of Chemical Physics, 2005. **122**: p. 224706.
36. Tsochatzidis, N., et al., *Determination of velocity, size and concentration of ultrasonic cavitation bubbles by the phase-Doppler technique.* Chem. Eng. Sci., 2001. **56**(5): p. 1831-1840.
37. Didenko, Y. and K. Suslick, *The energy efficiency of formation of photons, radicals and ions during single-bubble cavitation.* Nature, 2002. **418**(6896): p. 394-397.
38. Dudley, J., *Mass transfer in bubble columns: a comparison of correlations.* Water Research, 1995. **29**(4): p. 1129-1138.

# TiO2 Particle Property Investigation Featuring Moving Bed Adsorption and Photo-Assisted Degradation of Azo Dyes

**David M. Follansbee**, Department of Chemical and Biological Engineering, Rensselaer Polytechnic Institute, Troy, NY

John Paccione, Environmental Health Sciences, SUNY at Albany, School of Public Health, Troy, NY

Lealon L. Martin, Chemical and Biological Engineering, Rensselaer Polytechnic Institute, Troy, NY

The use of photon-driven chemical decomposition of organic pollutants such as benzene is a potentially viable solution for water polishing applications. Various attempts at advanced oxidation process (AOP) have yielded inherent operation and design limitations. To overcome these mass transfer, cross sectional area, and light scattering limitations a novel water treatment process has been developed relying on the characteristics of a circulating fluidized bed (CFB) system. This system allows for adsorption of a target contaminant onto the surface of a photocatalyst composite particle in a moving packed annular bed and photocatalytic degradation/regeneration of laden particles within a photon chamber in a continuous fashion.

The goal of this work is to independently study the adsorption and photocatalytic capabilities of photocatalyst-composite particles by effectively decoupling these two processes within the system. This study will give insight into the bottlenecks and limitations of photocatalyzed reactions by assisting in optimal design of these particles with a high quantum yield and efficient use of photons.

Azo dyes such as Procion Red and orange-II are used as model contaminants. TiO2 coated silica particles (.5mm, 1mm, and 2mm) are used as photocatalyst composite particles. An initial dye concentration of 500 ppm is used for all experiments. Photon chamber consists of a 2" OD quartz tube surrounded by six 50W UV bulbs (350 nm) in which the particles sediment through in lean phase fashion. Particles are removed prior to entering the UV chamber and upon exiting the UV chamber for analysis. Concentration of the contaminated water is monitored within the annular bed by a Beckman UV-Vis spectrophotometer.

See more of this Session: Advanced Oxidation/Reduction Applications: Liquid/Gas Phase II
See more of this Group/Topical: Environmental Division

# TiO2-BASED PHOTOCATALYSTS IN UV and VISIBLE REGION: Study of Activity and Stability

Aaron Ortiz-Gomez, Chemical and Biochemical Engineering, The University of Western Ontario, London, ON, Canada

Jesus Moreira, Chemical Engineering, University of Western Ontario, London, ON, Canada

Benito Serrano, Chemical Engineering, Universidad Autonoma de Zacatecas., Zacatecas Zacatecas., Mexico

Hugo H. De Lasa, University of Western Ontario, London, ON, Canada

Heterogeneous photocatalysis has emerged as a potential alternative for the degradation of organic pollutants in water. The TiO2 DP25 is the most active catalyst but it is limited to be used in the range of UV light. In order to make the DP25 active in the visible region, ion doping and coupling the DP25 with another photocatalysts have been tested. In this study, various catalysts were prepared using the sol-gel and wet impregnation techniques. Fe (III), Ni (III), and Ag (I) ions were used for the doping of DP25. AgBr was coupled with TiO2 using the deposition-precipitation method. Catalysts prepared by the sol-gel techniques and wet impregnation are tested in the UV region, while the AgBr/TiO2 catalysts are tested in the visible region. Both experiments were evaluated for the degradation of maleic acid in water. TiO2 + 5 ppm Fe catalyst has the higher activity in UV region but showed no activity in the visible region. On the other hand, AgBr/TiO2 catalyst was the most active in the visible region but showed lack of stability as it decomposed into Ag and Br2 while being irradiated. Thus, it is concluded that in order to utilize the AgBr/TiO2 catalyst under Vis radiation, it is necessary to find ways to make it stable under visible light. In this regard, it was found that AgBr/TiO2 catalyst could be stabilized for longer periods of time when the pH of the reacting solution was maintained at 2. A full characterization of the prepared catalysts is carried out to understand their structure and operational mechanism under the UV and Vis range to develop materials that are stable and active for longer reaction periods. This will allow the use of photocatalysts under both visible light and UV such as that naturally present in the sunlight.

See more of this Session: Advanced Oxidation/Reduction Applications: Liquid/Gas Phase II
See more of this Group/Topical: Environmental Division

# AUTHOR INDEX